T0262727

Encyclopedia of Alternative and Renewable Energy: Environment and Clean Energy

Volume 07

Encyclopedia of Alternative and Renewable Energy: Environment and Clean Energy
Volume 07

Edited by **Marrianne Fox and David McCartney**

New York

Published by Callisto Reference,
106 Park Avenue, Suite 200,
New York, NY 10016, USA
www.callistoreference.com

**Encyclopedia of Alternative and Renewable Energy:
Environment and Clean Energy: Volume 07**
Edited by Marrianne Fox and David McCartney

International Standard Book Number: 978-1-63239-181-0 (Hardback)

Printed in the United States of America.

Contents

Preface

This book has been a concerted effort by a group of academicians, researchers and scientists, who have contributed their research works for the realization of the book. This book has materialized in the wake of emerging advancements and innovations in this field. Therefore, the need of the hour was to compile all the required researches and disseminate the knowledge to a broad spectrum of people comprising of students, researchers and specialists of the field.

Alternative energy has emerged as an area of interest worldwide. The vision of producing fresh, sustainable power from renewable energy sources is becoming a significant subject. This is accelerated by current technological advancements that have enhanced the cost-effectiveness of many renewable sources and by the rising concern for the ecological impact and sustainability of conventional fossil and nuclear fuels. This book presents a complete comprehension of the primary renewable energy sources with a broad variety of case studies for each source. It explains the physical and technical facets, and examines the ecological impact of renewable sources and their potential prospects.

At the end of the preface, I would like to thank the authors for their brilliant chapters and the publisher for guiding us all-through the making of the book till its final stage. Also, I would like to thank my family for providing the support and encouragement throughout my academic career and research projects.

Editor

Grid-Connected Wind Park with Combined Use of Battery and EDLC Energy Storages

Guohong Wu, Yutaka Yoshida and
Tamotsu Minakawa

Additional information is available at the end of the chapter

1. Introduction

Over the last decade, driven by limited fossil fuel supply, global warming, and the provision of tax credit or financial support for renewable power production, sustainable power generations such as wind power and photovoltaic power etc. have been increasingly integrated into the existing power grids. Among these renewable power generations, wind power has various advantages such as large per unit capacity and easily construction of a large-scale generating station, which lead to a relatively lower generation cost over the other ones, and hence be considered as one of the mature renewable energy alternatives to the conventional fuel-based resources [1]. Fig. 1 and Fig. 2 gives the change tendency of the total introduction capacity and annual introduction capacity of wind power generation in the world [2]. It can be known from these figures that the wind power generation has been significantly increased in the recent years and is expected to increase further in the later future.

However, the main challenge of wind power utilization is associated with the fluctuation and unpredictability in its power generation. Fig. 3 gives a typical wind power output curve based on the measurement data [3]. This data is a good example to obviously indicate the facts that the wind power consists of both considerable fast/short-term fluctuation and slow/long-term variation. Generally, wind parks are located in the remote area and interconnected into the distribution system via a relatively long tie transmission line. Along with the increasing integration of wind power into a power grid, due to the fluctuation of wind power, considerable fluctuation of power flow in the tie transmission line (the line that connects the wind park to the power grid) may occurs and lead to some problems such as power quality degradation, voltage instability and insufficient available power transferability, etc., and hence imposes difficulties both in terms of operation and planning. These issues need to be resolved adequately to reduce the negative effects on the existing power grid in order to facilitate the increasing integration of wind power into the future power networks.

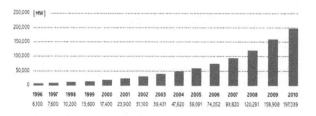

Figure 1. Change of the total introduction capacity of wind power generation in the world [2]

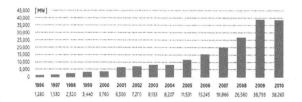

Figure 2. Change of the annual introduction capacity of wind power generation in the world [2]

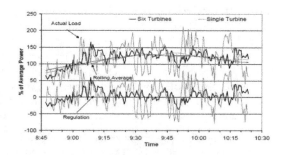

Figure 3. An example of actual power output of a wind plant and a single wind turbine generator [3]

For this purpose, in this chapter, it is considered that energy storage is an effective way for power management and power leveling for a wind park, especially in the case that wind power with significant fluctuation (e.g., at a wind park located in a mountainous region) is integrated into a distribution system with initially weak stability.

The concepts presented in this chapter mainly pay attentions to the power leveling and fluctuation mitigation effect by introduction of energy storages. A method that considers the combined use of two types of energy storage devices with different response properties and costs are proposed for a grid-connected wind park. These considered energy storages include:

(1) A secondary electrical battery, which is characterized as relatively slow response and low cost, is used with large capacity for power leveling of slow wind power variation.

(2) EDLC (Electrical Double Layer Capacitor), which has an extremely high response and long life cycle but is high cost, is introduced with small capacity for mitigating fast wind power fluctuation only.

This chapter presents the detailed models of EDLC and battery, together with the related control systems for mitigation of wind power fluctuations. For examining the improving effect by the combined use of these energy storage devices, digital simulations with a typical 66kV class distribution system model integrated by a wind park are conducted, the simulation results have illustrated the validity of the proposed method.

2. An overview of approaches to wind power fluctuation mitigation

Up to date, some approaches have been made to coping with these fluctuation problems caused by wind power integration. The idea of these studies can be simply divided into two types:

A. To mitigate the power fluctuation from each single wind turbine generator in a wind park

B. To mitigate the fluctuation of total power from a wind park that consists of a group of wind turbine generators

A brief summary of these two methods will be addressed in the following sessions.

2.1. Method Type (A) for mitigation of wind power fluctuation

Method Type (A) aims at mitigating the power fluctuation of each single generator in a wind park and by this way to obtain a stable power output from the whole wind park. The main idea of this method is to develop new type wind turbine generators by application of power electronic technologies, and improve the operation property of the wind turbine generator unit.

At the early stage of wind power development, most of the wind turbine generators are induction machine due to its simple structure, low cost and easily for maintenance; however, there are some problems with this type of generator as well, such as difficulty in control of power output, inrush current and FRT (Fault Right Through) problems. In order to treat these issues, some approaches have considered using new type of wind turbine generators instead of the induction ones. Typical example of these new type generators is Doubly-Fed Induction Generators (DFIG) or Variable Speed Generators (VSG) [4]-[6]. The basic configuration of DFIG is shown in Fig. 4. By adding a back to back PWM converter to the excitation circuit of the wind turbine generator, the active and reactive power of generator can be dynamically controlled, and hence, a stable power output can be obtained. Another approach is to connect the wind turbine generator to the power gird via an AC-DC-AC circuit (DC link) [7], since active power of the inverter with DC link can be properly controlled, the wind power penetrating into the grid is possibly stabilized as well. The basic configuration of wind turbine generator with DC link is given in Fig.5. These methods have been verified to be effective in mitigating fast fluctuation and FRT problems [4]~[7].

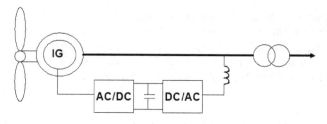

Figure 4. Basic configuration of a Doubly-Fed Induction Generator

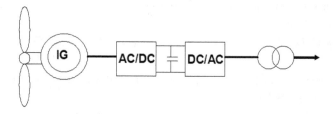

Figure 5. Basic configuration of a wind turbine generator with DC link

Method of Type (A) has the following properties:

- Because each of the generators has the ability to stabilize its own power output, there are not strict limits in the capacity or number of new wind power plants being added to the existing wind park.

- Compared with the conventional induction generators, these improved ones have few in-rush current during starting, and are slightly influenced by the variation of voltage or power at the wind park terminal bus.

- However, due to the necessity of power electronic circuits, these types of generator are normally higher cost than the conventional induction generator, and there is also consid-erable power loss in these power electronic circuits. In addition, extremely complicated control systems are necessary as well.

- Since there is not energy buffer ability as an energy storage device can provide, even the fast wind power fluctuation can be effectively suppressed by properly control of the pow-er electronic devices, considerable slow fluctuation may still penetrate into the power grid intermittently.

The power fluctuation may aggravate the power quantity or make it difficult for power management. And therefore, wind power integrations are strictly required to satisfy the grid-interconnect guideline.

2.2. Method Type (B) for mitigation of wind power fluctuation

Method Type (B) emphasizes the necessity of stabilizing and managing the total power output from the wind park. The main idea of this method is to control the power flow in the tie transmission line that connecting the wind park to the grid by use of some energy compensation systems such as a battery station, a STATCOM/BESS, a flywheel or SMES (Superconducting Magnetic Energy Storage) system, etc.

The energy storage device is generally installed at the bus near to the tie transmission line, and connected to the bus via a DC/AC inverter. Controls for these energy storages are designed to cancel out the active power fluctuation caused by wind power. Furthermore, power management (large amount of power charge/discharge responding to the power demand or request from the power grid) can be also performed. The basic concept of the method Type (B) is simply illustrated in Fig.6.

The introduction of properly controlled energy storages are confirmed to be able to effectively compensate most of the slow wind power fluctuation and reduce negative impacts on the existing power grid[8]-[15], in addition with the ability to bring about economic benefits by power management[16].

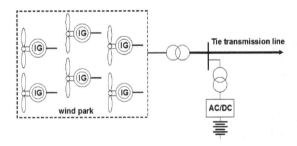

Figure 6. Basic concept of method Type (B) for mitigation of wind power fluctuation

Method Type (B) has the following properties:

- Only limited number of energy storage stations is necessary. It is space-saving and can be built compactly.

- Large capacity of power energy can be saved or released. Hence, the energy storage station can be employed not only as a method to compensate the slow wind power fluctuation that is accompanied with large change of power energy, but also as a meaning to support the power demand-supply regulation (charge or discharge at required time duration) in the grid.

- Since the power flowing out the wind park can be stabilized, it is unnecessary to mitigate the power output for every single wind generator in the wind park, and therefore, most of the generators can possibly use the low cost induction machines.

- However, in case that some new wind turbine generators are to be added to the existing wind park, the necessary capacity of energy storage may change and it is considered not an easy work to vary the capacity of an existing energy storage station.

- Beside, the high cost of energy storage devices is also a big issue. Presently, some new energy storage devices of low cost, high efficiency and high energy density are under development.

3. An overview of energy storages application for wind power systems

In the recent year, along with the increasing utilization of renewable energy sources and the remarkable advance in smart grid technology, the energy storage is considered as one of the key devices in the next generation power networks. Among the art-of-energy storage, beside the material and production technologies, the important issues of energy storage application in power systems are considered as the followings:

- Design of control systems for mitigation of power fluctuation and power management

- Determination of adequate capacity of energy storage devices and power converters

- Identification of proper location of energy storage stations in a power grid

Up to date, many studies have been done related to the application of energy storages. The works in (8) - (10) discussed the design of control systems and the determination of optimal capacity of battery energy storages for the purpose of wind power stabilization, and have illustrated the improving effect by introduction of batteries in a wind park. Reference (11) - (13) presented an approach to the application of EDLC (Electric Double Layer Capacitor, referred to as Super-Capacitor as well) in a wind power system. The results from these works have shown the effectiveness of EDLC for mitigation of extremely fast wind power fluctuation. Paper (14) studied the wind power stability improved by a SMES (Superconducting Magnetic Energy Storage) system. Reference (15) advocated the combined use of EDLC and battery for suppressing both the fast and slow wind power fluctuation, and (16) investigated the operation and sizing of energy storage in terms of economic benefits in the power market.

4. Why combined use of energy storages is necessary for a wind power system?

Large-scale wind parks may be built on plains, offshore and in mountainous regions. For wind parks located offshore or on plains where stable, strong wind conditions are available, fast fluctuation of wind power may not be a serious problem since in these cases, naturally few fast power fluctuations occur, furthermore, such fluctuations are smoothened by the power leveling effect with different wind turbine generators distributed over a broad area. If these wind parks do not directly supply power to local loads or are not be interconnected to an extremely weak power grid, energy storage is even unnecessary.

However, in case of wind parks constructed in complicated topographical mountainous regions where good wind conditions can hardly be obtained, significant and frequent changes of wind speed and wind direction may occur, and both extremely fast fluctuations and slow variations appear with wind power output, energy storage thus be necessary for dealing with the problems caused by these fluctuations. Furthermore, if only a secondary electrical battery is used, the slow response of battery make it hardly compensate the extremely fast power fluctuation, and in addition, the battery may possibly charge or discharge frequently responding to both the fast and slow changes of wind power, resulting in significant shortening of its service time. In such case, the combined use of an EDLC and a battery is considered to be an effective choice.

5. The idea of combined use of different types of energy storages for a wind power system

The idea of combined use of EDLC and battery in a wind park is classified to the method of Type (B) in section 2. It aims at obtaining both mitigation effects of fast power fluctuation and power leveling effect of slow power variation in the tie transmission line, and in the meantime, avoiding frequent charge/discharge operation of battery. The basic concept of a wind park with combined use of EDLC and battery is shown in Fig.7.

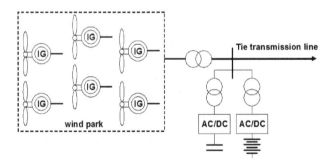

Figure 7. Basic concept of a wind park with combined use of EDLC and battery

Taking consideration of the facts that EDLC has extremely high response, long life cycle but with low energy density and high cost, whereas battery is relatively low cost but with slow response and short life cycle, EDLC is employed as a manner to mitigate the fast fluctuations in a small capacity, and battery is used as the one to deal with the slow variation with large capacity.

6. Introduction and analysis model of EDLC

6.1. About EDLC

EDLC is a newly developed electrical storage device which has recently attracted great attentions. This is kind of electro-chemical capacitor that has large energy storage capacity,
and therefore often called a super- or ultra-capacitor. Since EDLC has several advantages
such as an extremely long life-cycle, no contamination, operation under normal temperature
and extremely rapid charge/discharge operation from 0% to full capacity with less voltage
loss, it is considered suitable for use in the case when frequent charge/discharge is necessary. Nevertheless, it also has disadvantages such as low energy density and high cost, and
thus, it is not preferable to use EDLC as a manner for large-capacity energy storage.

EDLC modules with capacities from several to several thousands Farads are already commercially available. For power energy storage such as that used in this study, an EDLC bank
needs to be made by means of connecting several hundreds of EDLC cell units in parallel
and in series.

6.2. Analysis model of EDLC

Since the EDLC is a newly developed device and still under study, there is no ready-made
model for conducting simulations. In this work, an analysis model of EDLC is derived based
on an equivalent circuit presented in reference [17].

Electrical characteristic expression of EDLC can be basically expressed in the form of

$$V_S(s) = I_S(s) \cdot f(s) + Q_{C0} \cdot g(s) \tag{1}$$

Where, $V_s(s)$ is the DC voltage, $I_s(s)$ is the DC current and Q_{C0} is the initial charged capacity
of the EDLC; $f(s)$ and $g(s)$ are transfer functions from $I_s(s)$ to $V_s(s)$ and that from Q_{C0} to $V_s(s)$,
respectively.

In expression (1), $f(s)$ and $g(s)$ need to be derived. According to the work in [17], an EDLC
bank can be electrically expressed by an equivalent circuit as shown in Fig.8, which takes the
allotment of internal resistance, leak resistance and capacitance of an EDLC unit into account and is referred to as "Double-Layer Equivalent Circuit." In Fig.8, R_s and R_p are internal resistance, and R_L is leakage resistance. C_1 and C_2 are equivalent electrostatic
capacitances, respectively.

Based on Fig.8 and applying electrical circuit theory, the electrical characteristic of EDLC
can be obtained as expression (2).

$$V_s(s) = I_s(s) \left\{ \frac{C_2 R_p R_L\ s + R_L}{C_1 C_2 R_p R_L\ s^2 + (C_1 R_L + C_2 R_L + C_2 R_p)s + 1} + R_s \right\}$$

$$+ \frac{C_2 R_p R_L\ s + R_L}{C_1 C_2 R_p R_L\ s^2 + (C_1 R_L + C_2 R_L + C_2 R_p)s + 1} Q_{10} \qquad (2)$$

$$+ \frac{R_L}{C_1 C_2 R_p s^2 + (C_1 + C_2 R_L + C_2 R_p)s + 1} Q_{20}$$

Herein, Q_{10} and Q_{20} represent the initial charged capacities of C_1 and C_2 in Fig.8.

Expression (2) gives the electrical characteristics of the DC voltage V_s responding to the change of DC current I_s and can be used as an analysis model of EDLC for simulation study.

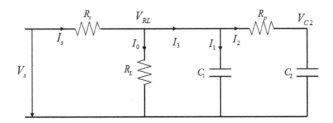

Figure 8. Double-layer equivalent circuit model for EDLC bank

7. Introduction and analysis model of battery

7.1. About battery

Batteries are the most widely used energy storage devices. At present, several types of secondary electrical batteries have been developed or are under development. Among these batteries, the nickel-metal hydride battery and the lithium-ion battery are popular types for consumer electronics (portable electronics with small capacity) and are still under development. The sodium-sulfur (NaS) battery is a new type of molten metal battery and primarily suitable for large-scale non-mobile applications. And the lead acid battery charges/discharges by utilization of chemical reaction with the metallic lead soaked in a solution of diluted sulphuric acid. Being fabricated from inexpensive materials, it is of relatively low cost. In addition, it can operate in normal temperature. Given these advantages, this type of battery is already widely used in various fields and mass-production technology has been completely established. Furthermore, its operational characteristics have been fully elucidated. For this reason, a lead acid battery was selected for large-scale energy storage in this study. However, because the dynamics of this type of battery are governed by chemical reactions, its response is not as quick and its cycle life is not as long as those of an EDLC. For this reason, the lead acid battery is considered to be suitable to deal with slow wind power fluctuation with relatively large capacity so as to reduce the usage of the high-cost EDLC.

For obtaining large capacity, the battery bank consists of large quantity of cell units connecting in parallel and in series.

7.2. Analysis model of battery

An equivalent circuit of Fig.9 is used for deriving the analysis model of battery [18].

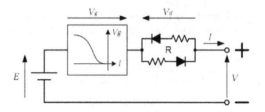

Figure 9. Equivalent circuit model of lead acid battery

According to Fig. 9, the terminal voltage V of the battery can be calculated by

$$V = E - V_d + V_g \tag{3}$$

Where, E is the equivalent electromotive force, V_d is the voltage loss in circuit resistance R, and V_g is the additional voltage increase at time toward the end of charge. These variables are calculated by the following expressions:

$$E = E_0 - \Delta E(1 - SOC) \tag{4}$$

$$V_d = RI + V_a(1 - e^{-\frac{|I|}{K_a}}) \tag{5}$$

$$V_g = \begin{cases} 0 \quad during discharge \\ \dfrac{V_{gmax}}{2} e^{\frac{SOC-SOC_g}{K_g|I|}} \quad whiheSOCSOC_g during charge \\ \dfrac{V_{gmax}}{2}(2 - e^{\frac{SOC-SOC_g}{K_g|I|}}) \quad whiheSOC \geq SOC_g during charge \end{cases} \tag{6}$$

Herein, SOC is the state of capacity, E_0 is the value of E in the fully charged condition, ΔE is the change of E responding to SOC, V_a is the maximum value of the nonlinear component of V_d, V_{gmax} is the maximum value of V_g and SOC_g is the capacity where the additional voltage starts to increase during the charge, and K_a and K_g are coefficients.

Expression (3) – (6) can be used as an analysis model of battery for simulation study.

8. Charge and discharge strategy for energy storages

EDLC may physically charge and discharge to its full capacity, however, whereas working with a DC/AC inverter for connecting to the AC power grid, extremely low DC voltage may lead to malfunctioning of the inverter. On the other hand, discharge of the battery to the level under half of its capacity may significantly shorten its life cycle. For this reason, a low limit for DC voltage of EDLC and battery bank is considered to ensure that DC voltage does not decrease to an extremely low level during its discharge process. On the other hand, DC voltage must not extend beyond the rated voltage value of the energy storage device so as to protect it from being destroyed. Thus, an upper limit for DC voltage is necessary as well.

This process is obtained by the follow strategy:

- During the process in discharge mode, while DC voltage reaches a low limit, discharge is stopped automatically and energy storage device automatically switches to the charge mode.

- During the process in charge mode, while DC voltage approaches an upper limit, charge is stopped automatically and the energy storage device is ready to discharge.

- In this study, the low limit and upper limit were set to be 0.5 p.u. and 0.9 p.u., respectively.

Furthermore, in order to avoid the phenomenon that discharge or charge operation is repeated around the limit boundary, the following requirement is added in the charge/discharge control:

- Whenever the EDLC changes its charge or discharge mode, this mode must be continued until the DC voltage reaches 0.7 p.u.

Based on the above control strategy, the flow chart of the EDLC charge/discharge control as depicted in Fig.10 is proposed.

9. Wind power generation system model with introduction of combined use of energy storages

In this chapter, in order to examine the validity of the combined use of EDLC and battery energy storages, a typical wind power system model shown in Fig. 11 is created for conducting simulation study. In this model, the wind park is assumed to have a total capacity of 30 MW and is simulated by three induction generators, each of which has a capacity of 10 MW representing equivalently a cluster of wind turbines generators (e.g., 4 generators of 2.5 MW class) connected to a 22 kV bus. Generated power from the wind park is supplied to local loads and the remaining portion is sent to the infinite bus (bulk system) through tie transmission line 3.

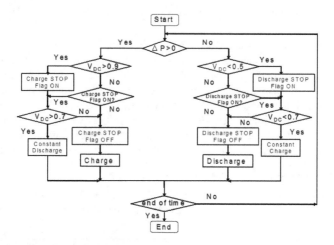

Figure 10. Flowchart of charge/discharge control for EDLC and battery

EDLC and battery energy storage systems are introduced at the substation in Bus 4 between the tie line and the wind park. Both of these energy storage systems are connected to Bus 4 via a DC/AC inverter with a capacity of 5.0 MW.

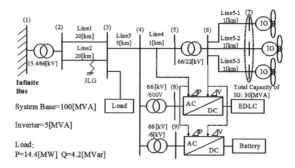

Figure 11. Wind power system model with combines use of EDLC and battery

The simulation study was completed by use of a power system analysis software package named "MidFielder," which is developed by TEPCO (Tokyo Electric Power Co., Japan). Transmission line and transformer model used the default ones with parameters prepared by the software package, which is created based on the real data of power network in Japan.

9.1. Wind Turbine Generator Model

Taking introduction costs of the wind park into account, all of the wind turbine genera-tors are assumed to be low-cost induction generators. The induction generator model used the standard model with parameters prepared by the software package, which are list-ed in Table 1.

9.2. Parameters of EDLC and battery

Parameters of the EDLC bank and the battery bank which appear in Fig.8 and Fig.9 are tabu-lated in Table 2 and Table 3, respectively. Comparing the capacity of the EDLC and battery in these tables, it is obviously seen that the capacity of the battery is considerably larger than that of the EDLC. Working with these capacities, the EDLC can continue charge/discharge at time duration of 20 sec. with inverter outputting in full capacity, and the battery extends the time to 6 hours (assuming 50% of capacity can be used for energy storage).

Constants	Values
Pole Number	4
Stator resistance (p.u.)	0.002
Stator leakage reactance (p.u.)	0.11
Rotor resistance (p.u.)	0.013
Rotor leakage reactance (p.u.)	0.12
Excitation reactance (p.u.)	3.9
Inertia constant (s)	1.5
Rated slip	0.012

Table 1. Parameters of induction generators

Parameters	Values	Parameters	Values
Capacity[MJ]	388.8	$R_s[m\Omega]$	0.089
$C_1[F]$	1512	$R_p[m\Omega]$	0.837
$C_2[F]$	648	$R_i[k\Omega]$	3.934

Table 2. Parameters of EDLC bank

9.3. Inverter model and control systems

A DC/AC inverter is necessary for connecting the EDLC and battery bank to the AC power grid. The inverter model in this simulation study used the standard one prepared by the software package, which is an ideal inverter model ignoring power loss and represents its electrical characteristics by use of analysis models including a conversion transformer model, a 6-bridge inverter circuit model, a PLL model, a 3 phase–dq transformation model and a current control PWM model.

Parameters	At Charge	At Discharge
Capacity	2000Ah(2.16x10 5MJ)	
E_0 [kV]	6.621	6.423
ΔE[kV]	0.726	0.645
V_{gmax} [kV]	1.236	-
$R[m\Omega]$	0.863	0.261
K_a	213.143	213.143
K_g	0.0003	-

Table 3. Parameters of Battery bank

The active power and reactive power output from the inverter model can be controlled. Active power output is controlled by I_{dinv} (AC current in d axis of coordinates) and reactive power is outputted responding to I_{qinv} (AC current in q axis of coordinates). The control for I_{dinv} and I_{qinv} is designed in this study. The control block for DC/AC inverter connected to the EDLC is shown in Fig. 12 and that for the battery is given in Fig.13.

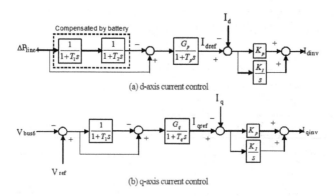

(a) d-axis current control

(b) q-axis current control

Figure 12. Control blocks for EDLC inverter

In Fig.12 and Fig.13, ΔP_{line4} is the deviation of active power in transmission line 4; V_{bus6} and V_{ref} are the measured and reference value of Bus 6 voltage, respectively; and I_d and I_q are measured AC current in d-q axis of coordinates. Control parameters in these control blocks are set as follows:

G_p=20, G_q=66, K_p=0.4, K_I=10, and T_p=T_q=0.04[sec], T_1= 2.0[sec], T_2=5.0[sec]

It can be known from these control blocks that in order to level the power flow in tie transmission line 3, so as that the power fluctuation from wind park does not affect the local load and bulk system, the deviation of power in line 4 (the transmission line between the wind park and the tie transmission line 3) is used to control the active power output of the EDLC and battery; Meanwhile, in order to mitigate the voltage fluctuation in Bus 4 where the wind park is integrated into the distribution system, voltage deviation of wind park terminal Bus 6 is adopted as a control signal to control reactive power from these energy storage devices.

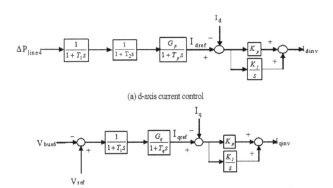

(a) d-axis current control

Figure 13. Control blocks for battery inverter

These control systems are designed to control the inverter output cooperatively. The cooperative control strategy depicted in Fig.12 and Fig.13 is explained in the following.

(a) The deviation of active power of line 4 is calculated and used as the control signal for I_d current so that the active power output from energy storage devices can cancel out the power variation of line 4 and thus level the active power flowing into tie transmission line 3.

(b) Voltage deviation of Bus 6 is used to control I_q current so as that these energy storage devices can output adequate reactive power to suppress the voltage deviation of Bus 6 from its reference.

(c) Slow power variation is compensated for by the battery, and this is realized by adding low-pass filters with a time constant of T_1 to the control block for the battery. Meanwhile, the fast fluctuation is mitigated by the EDLC, which is calculated by excluding the compensated portion by battery from the input signal.

10. Simulation studies

10.1. Wind speed data

For simulation study, wind speed data given in Fig.14 are modified based on the actual wind velocity data measured with an anemometer at the campus of Tohoku Gakuin University, Japan. In this study, wind data of 1000 seconds measured with time-interval of 1.0 second are selected for simulation.

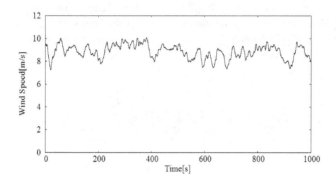

Figure 14. Wind speed data

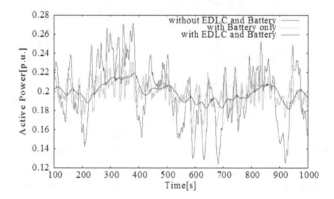

Figure 15. Active power of tie transmission line 3

10.2. Simulation results and comments

Based on the above-mentioned simulation conditions, simulation studies were conducted to verify the following properties:

- Mitigation effect of fast and slow fluctuations of power flow in tie transmission line 3
- Enhancement effect of voltage stability of Bus 4 where the wind park is integrated into the distribution grid
- Verification of FRT capability

Mitigation effect of power flow in tie transmission line

With the wind turbine generators driven by the fluctuating wind speed in Fig.14, the power flow in tie transmission line 3 in cases of "without energy storage," "with battery only" and "with both EDLC and battery" are shown in Fig.15, and the active power output from the EDLC and battery are given in Fig. 16.

Comparing the change of active power curves in Fig.15, it can be seen that the active power of tie line 3 fluctuates considerably in the case without any energy storage devices, whereas in the case with only a battery, although slow variation of power flow is compensated, some fast fluctuation still appears. This result indicates that the slow response of the battery result in its difficulty to trace and absorb the extremely fast fluctuation. In contrast, it can be verified that in the case with both the EDLC and battery, power flow is effectively suppressed and leveled.

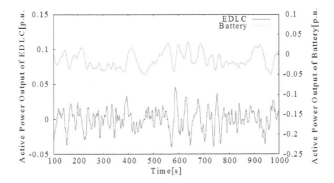

Figure 16. Active power output from EDLC and battery

From Fig.16, it can be confirmed that the active power from the EDLC is significantly changing so as to mitigate the fast fluctuation and that meanwhile, the power from the battery is slowly changing to compensate for the slow variation of the power flow.

Enhancement of bus voltage stability

Under the same simulation conditions, the wave curve of the terminal voltage of Bus 4, where wind power is integrated into the distribution system, is shown in Fig. 17. From Fig. 17, it can be seen that the voltage of Bus 4 fluctuates significantly from 0.9 [p.u.] to 0.984 [p.u.] if without any compensation method, whereas in cases with only the battery and with the EDLC and battery, the voltage is compensated to 0.995-1.002 [p.u.]. Furthermore, it is also known that even with only the battery, the voltage stability is enhanced remarkably, and the improving effect by the EDLC is not obviously observed.

The reactive power output from battery and EDLC system are given in Fig. 18. It can be known from this result that reactive power of the battery is obviously larger than that of the EDLC, which means that voltage variation is mainly compensated by the battery. In addition, rapid reactive power change of EDLC is observed due to the fast voltage fluctuation of

Bus 6 (wind park terminal bus). However, unlike the active power flow in the tie transmission line 3 in Fig. 15, which is directly affected by the active power from the wind park, the voltage of Bus 4 is influenced indirectly by the voltage of wind park terminal Bus 6. For this reason, although the EDLC can bring about a mitigation effect of fast voltage fluctuation in the voltage of Bus 6, this enhancement effect is not remarkably revealed in the voltage curve of Bus 4. This result also indicates that if only voltage stability enhancement in the interconnection bus 4 is required and the voltage feature of Bus 6 is considered being not significantly important, it is sufficient to add voltage compensation function only to the battery and EDLC devices may not be mandatory for this purpose.

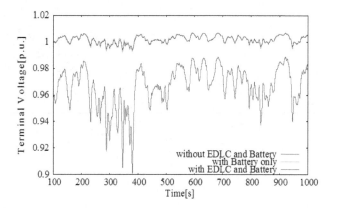

Figure 17. Voltage of bus 4 (the bus where wind park is interconnected to the distribution grid)

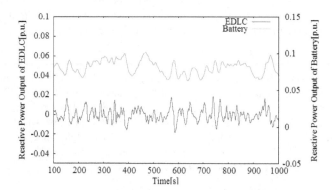

Figure 18. Reactive power output from battery and EDLC

Verification of FRT (Fault Ride Through) capability

Under the same simulation conditions and assuming that a 3LG fault (100 ms) occurs at 225 second at line 2 (near bus 3), the simulation result of wind park terminal voltage of Bus 6 is shown in Fig. 19. From this figure, it is known that in the case without any energy storage, because the dramatic decrease in bus voltage occurs along with the fault and without any voltage support, bus voltage can not recover after the fault, and wind turbine generators have to be tripped from the power grid to avoid accelerated step out. On the other hand, because the energy storage devices can supply reactive power simultaneously following the fault and facilitate the recovery of bus voltage, the introduction of the EDLC and battery leads to the rapid recovery of bus voltage after the fault, and wind turbine generators can continue operation. This is usually referred to as the FRT capability.

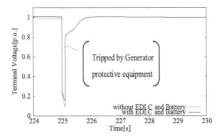

Figure 19. Voltage of Bus 6 (terminal bus of wind park)

11. Conclusions

In the recent year, along with the increasing concerns of global warming and exhaustion problem of fossil energy resources, sustainable power generations such as wind power and photovoltaic power etc. have been increasingly integrated into the existing power grids. However, some challenging issues associated with the fluctuation and unpredictability in renewable power generation have to be adequately treated so as to facilitate its further de-velopment in the future power networks. This chapter has presented one of the technologies for construction of a wind park with application of energy storages.

This chapter has given an overview of the current technologies for an improved wind power system, and then presented a simulation-based approach to dealing with the fluctuation problems in terms of power flow and voltage stabilizations, which is caused by wind power integration into a distribution system. And a method that takes consideration of combined use of energy storage devices with different response characteristics and costs was proposed and verified.

The concepts of this chapter are summarized in the following:

- For the purpose of mitigating the power fluctuation due to wind power generation, up to data, technologies applied to a single wind turbine generator such as DFIG or DC link was studied and some of these technologies have been already in practical use; on the other hand, technologies considering the total operation of a wind park such as utilization of centralized energy storages station is noticeably developed as well.

- From the point of view of wind power stabilization and power management, the energy storage is considered to be an extremely important technology for the stable operation of wind parks especially that are integrated to relatively weak distribution systems.

- The combined use of EDLC and battery is an advisable way for obtaining a satisfactory improved effect of power flow leveling and voltage stability for a wind power system, especially which located in complicated topographical conditions (e.g. the mountainous regions) where good wind conditions can hardly be obtained and with both fast and slow fluctuations in wind speed.

- In the combined energy storage system, a battery can be used in large capacity for leveling long-term power flow variation and an EDLC can be used to mitigate the short-term fluctuation with a relatively small capacity.

Author details

Guohong Wu[1*], Yutaka Yoshida[1] and Tamotsu Minakawa[2]

*Address all correspondence to:

1 Dept. of Electrical of Electrical Engineering & Information TechnologyTohoku Gakuin University, Japan

2 NPO ECA, Japan

References

[1] Technical Report of Japan Society for the Promotion of Science Research for the Future Program:. (2002). "Highly Efficient Use of Energy and Reduction of its Environmental Impact", 22-23.

[2] GWEC, "Global Wind 2010 Report". (2011). Global Wind Energy Council, April.

[3] Ernst, B., Wan, Y. H., & Kirby, B. (1999). "Short-Term Power Fluctuation of Wind Turbines: Analyzing Data from the German 250-MW Measurement Program from the Ancillary Services Viewpoint". *Proc. Windpower '99 Conference, Burlington, Vermont, USA, June.*

[4] Pena, R., Clare, J. C., & Asher, G. M. (May 1996). "Doubly fed induction generator using back-to-back PWM converters and its application to variable-speed wind-energy generation". *IEE Proc. Electric Power Applications*, 143, 231-241.

[5] Muller, S., Deicke, M., & De Doncker, R. W. (Jun.2002). "Doubly fed induction gener-ator systems for wind turbines". *IEEE Trans. Industry Applications*, 8, 26-33.

[6] Datta, R., & Ranganathan, V. T. (Sept. 2002). "Variable-speed wind power generation using doubly fed wound rotor induction machine - a comparison with alternative schemes". *IEEE Trans. Energy Conversion*, 17, 414-421.

[7] Song, S.H., Kang, S.i., & Hahm, N.k. (2003). "Implementation and control of grid con-nected AC-DC-AC power converter for variable speed wind energy conversion sys-tem". *Proc.Applied Power Electronics Conference and Exposition*, 1, 154-158.

[8] Teleke, S., Baran, M.E., Bhattacharya, S., & Huang, A.Q. "Optimal Control of Batter-yEnergy Storage for Wind Farm Dispatching". *IEEE Trans. Energy Conversion*, 25, 787-794.

[9] Black, M., & Strbac, G. "Value of Bulk Energy Storage for Managing Wind Power Fluctuations". *IEEE Trans. Energy Conversion*, 22, 197-205.

[10] Nanahara, T. (2009). Capacity Requirement for Battery Installed at a Wind Farm. *IEEJ Trans. on Power and Energy*, 129(5), 645-652.

[11] Kinjo, T., Senjyu, T., Urasak, N., & Fujita, H. "Output levelling of renewable energy by electric double-layer capacitor applied for energy storage system", *IEEE Trans. Energy Conversion*, 21, 221-227.

[12] Abbey, C., & Joos, G. (2007). "Supercapacitor Energy Storage for Wind Energy Appli-cations", *IEEE Trans. Industry Applications*, 43, 769-776.

[13] Goto, M., Wu, G., Tada, Y., & Minakawa, T. (2007). "The Basic Study of Influence of Energy Storage Systems on WindPower Generation Systems", *IEEJ Trans. Power and Energy*, 127(5), 637-644.

[14] Nomura, S., Ohata, Y., Hagita, T., Tsutsui, H., Tsuji, S., & Shimada, R. (2005). "Wind farms linked by SMES systems", *IEEE Trans. Applied Superconductivity*, 15, 1951-1954.

[15] Yoshida, Y., Wu, G., & Minakawa, T. "Mitigating Fluctuation of Wind Turbine Power Generation System by Introduction of Electrical Energy Storage with Combined Use of EDLC and Lead Acid Storage Battery", *Proc.POWERCON2010, HangZhou, China*, (FP0689).

[16] Korpaasa, M., Holena, A. T., & Hildrum, R. (2003). "Operation and sizing of energy storage for wind power plants in a market system", *Journal of Electrical Power and En-ergy Systems*, vol.25, 599-606.

[17] Okamura, M. (2000). Electric Double Layer Capacitor and Energy Storage System, *Nikkan Industrial Publication Corp.*

[18] Shimada, T., Kurokawa, K., & Yoshioka, T. "Highly Accurate Simulation Model of Battery Characteristics". *Conf. Rec. IEEJ Annual Meeting*, 7, 48-49.

Chapter 2

Exergy Analysis of 1.2 kW Nexa™ Fuel Cell Module

G. Sevjidsuren, E. Uyanga, B. Bumaa, E. Temujin,
P. Altantsog and D. Sangaa

Additional information is available at the end of the chapter

1. Introduction

There are two key problems with continued use of fossil fuels, which provide about 80% of the world energy demand today. The first problem is that they are limited in amount and sooner or later depleted. The second problem is that fossil fuels are causing serious environmental problems, such as global warming climate changes.

The fuel cell technology is friendly energy conversion with a high potential for environmentally. Fuel cells are ideally suited for applications that require electrical energy as the end. Fuel cell systems operate at higher thermodynamic efficiency than heat engines and turbines.

The fuel cell converts chemical energy directly into electricity by combining oxygen from the air with hydrogen gas without combustion. If pure hydrogen is used, the only material output is water and almost no pollutants are produced. Very low levels of nitrogen oxides are emitted, but usually in the undetectable range. The hydrogen can be produced from water using renewable energy forms like solar, wind, hydro or geothermal energy. Hydrogen also can be extracted from hydrocarbons, including gasoline, natural gas, biomass, landfill gas, methanol, ethanol, methane and coal-based gas.

Today, practical fuel cell systems are becoming available and are expected to attract a growing share of the markets for automotive power and generation equipment as costs decrease to competitive levels. Depending on the type of fuel cells, stationary applications include small residential, medium-sized cogeneration or large power plant applications. In the mobile sector particularly low-temperature fuel cells, can be used for passenger vehicles, trains, boats, and air planes [1-2].

Proton Exchange Membrane (PEM) fuel cell: Proton Exchange Membrane (PEM) fuel cells are currently the most promising type of fuel cell for automotive use and have been used in the majority of prototypes built to date. PEM fuel cells (membrane or solid polymer) typically operated at relatively low temperatures (~50-100°C), have high power densities, can vary their output quickly to meet shifts in power demand, and are suited for many applications.

PEM fuel cells used an electrolyte such as conducted hydrogen ions from the anode to cathode. The electrolyte is composed of a solid polymer film that consists of a form acidified Nafion membrane. The membrane is coated on both sides with highly dispersed metal alloy particles (mostly platinum or platinum alloys) that are active catalysts. Hydrogen is fed to the anode side of the fuel cell where, due to the effect of the catalyst, hydrogen atoms release electrons and become hydrogen ions (protons). The electrons travel in the form of an electric current that can be utilized before it returns to the cathode side of the fuel cell where oxygen is fed. The protons diffuse through the membrane to the cathode, where the hydrogen atom is recombined and reacted with oxygen to produce water, thus completing the overall process [3].

In this work, we will focus on the efficiency of a PEM fuel cell system at variable operating conditions such as working temperature, pressure and air stoichiometry. Determination of an effective utilization of a PEM fuel cell and measuring its true performance based on thermodynamic laws are considered to be extremely essential. Thus, it would be very desirable to have a property to enable us to determine the work potential of a given amount of energy at power plant. This property is exergy, which is also called the availability or available energy.

In an energy analysis, based on the first law of thermodynamics, all forms of energy are considered to be equivalent. The loss of quality of energy is not taken into account.

An exergy analysis, based on the first and second law of thermodynamics, shows the thermodynamic imperfection of a process, including all quality losses of materials and energy, including the one just described. An energy balance is always closed as stated in the first law of thermodynamics. There can never be an energy loss, only energy transfer to the environment in which case it is useless. From the second law of thermodynamics, the exergy analysis of the irreversibility of a process due to increase in entropy. Exergy is always destroyed when a process involves a temperature change. This destruction is proportional to the entropy increase of the system together with its surroundings. Therefore, exergy is a property of the system–environment combination and not of the system alone.

Theoretically, the efficiency of a PEM fuel cell based on the first law of thermodynamics makes no reference to the best possible performance of the fuel cell, and thus, it could be misleading. On the other hand, the second law efficiency or exergetic efficiency of a PEM fuel cell, which is the ratio of the electrical output over the maximum possible work output, could give a true measure of the PEM fuel cell performance. Energy analysis performed on a system based on the second law of thermodynamics is known as exergy analysis [4-8].

2. Exergy analysis of 1.2kW Nexa™ PEM fuel cell

Exergy analysis is a thermodynamic analysis technique based on the second law of thermodynamics, considering of all components and parametric in the system.

In particular, exergy analysis yields efficiencies which provide a true measure of how nearly actual performance approaches the ideal, and identifies more clearly than energy analysis the causes and locations of thermodynamic losses. Consequently, exergy analysis can assist in improving and optimizing designs. A main aim of exergy analysis is to identify exergy efficiencies and the causes of exergy losses. The exergy of a system is defined as the maximum shaft work that can be done by the composite of the system and a specified reference environment. Typically, the environment is specified by stating its temperature, pressure and chemical composition.

Exergy efficiency: Exergetic efficiency, which is defined as the second law efficiency, gives the true value of the performance of an energy system from the thermodynamic viewpoint. The exergy efficiency of a fuel cell system is the ratio of the electrical output power and actual exergy.

Actual exergy defined as difference between the exergy of the reactants (hydrogen + air) and the exergy of the products (air + hydrogen). In the PEMFC module, a basic reaction occurs as below.

$$H_2 + Air \rightarrow$$
$$H_2O + Unused Air(Oxygen depleted air) + \qquad\qquad (1)$$
$$Electrical Power + Heat$$

The exergy efficiency of a fuel cell system is the ratio of the power output, over the exergy of the reactants (hydrogen + air), which can be determined [9-11] by following formula:

$$\eta_{exergy\ system} = \frac{Electrical\ output\ power}{Actual\ exergy}$$
$$\eta_{exergy\ system} = \frac{Electrical\ output\ power}{(Exergy)_R - (Exergy)_P} = \qquad\qquad (2)$$
$$\frac{\dot{W}_{elect}}{(\dot{E}_{air,R} + \dot{E}_{H_2,R}) - (\dot{E}_{air,P} + \dot{E}_{H_2O,P})}$$

where: $\dot{W}_{elect}$ – *electrical output power [kW]; $\dot{E}_{air,R}$, $\dot{E}_{H_2,R}$, $\dot{E}_{air,P}$, $\dot{E}_{H_2O,P}$ – total exergies of the reactants [kW]; air and hydrogen, and the products air and water, respectively.*

Electrical output power: The electrical power (gross power) production is the sum of the parasitic load (i.e. the NEXA™ blower, compressor, and control system load) and the external load (e.g. residential load).

$$\dot{W}_{elect} = \dot{W}_{para} + \dot{W}_{net} \qquad\qquad (3)$$

The external load ($\dot{W}_{net}$-net power) is calculated directly from the voltage and current measured at the load.

$$\dot{W}_{net} = I_{net} \cdot V_{net} \tag{4}$$

where: I_{net} -current measured at the external load (A) , V_{net} -voltage measured at the external load (V) .

Assuming all hydrogen is reacted, for every mole of hydrogen consumed, two moles of electrons become available. Using Faraday's constant (F), the mass flow rate of hydrogen $\dot{m}_{H_2}$ and the number of cells in the stack, a theoretical current for the NEXA™ power module can be found using following equation:

$$I_{Nexa} = \frac{2 \cdot F \cdot \dot{m}_{H_2}}{47} \tag{5}$$

The total theoretical electrical power output of the Nexa can be computed using the voltage of the stack (V_{Nexa}).

$$\dot{W}_{elect} = I_{Nexa} \cdot V_{Nexa} \tag{6}$$

Parasitic loads are estimated as the difference between the primary load and the theoretical electrical power calculated from fuel consumption because power consumption by the individual NEXA™ Sub systems was not measured.

Actual exergy: We already know it before, that the actual exergy is a difference between the exergy of the reactants and the exergy of the products. So intend to calculate the actual exergy we must to know the sub exergies (total exergy). The total exergy of the reactants and the products can be determined through the following equations:

$$\dot{E}_{air,R} = \dot{m}_{air,R}\left(ex_{ch} + ex_{ph}\right)_{air,R} \tag{7}$$

$$\dot{E}_{H_2,R} = \dot{m}_{H_2,R}\left(ex_{ch} + ex_{ph}\right)_{H_2,R} \tag{8}$$

$$\dot{E}_{air,P} = \dot{m}_{air,P}\left(ex_{ch} + ex_{ph}\right)_{air,P} \tag{9}$$

$$\dot{E}_{H_2O,P} = \dot{m}_{H_2O,P}\left(ex_{ch} + ex_{ph}\right)_{H_2O,P} \tag{10}$$

where: ex_{ph} - *physical exergy* $[kJ/kg]$; ex_{ch} - *chemical exergy* $[kJ/kg]$; $\dot{m}$ − *mass flow rates of the reactants and products* $[kg/s]$.

Physical exergy: Physical exergy, known also as thermo mechanical exergy, is the work obtainable by taking the substance through reversible process from its initial state (T, P) to the state of the environment (T_o, P_o). The general expression of the physical exergy can be described as:

$$ex_{ph} = (H - H_o) - T_o(S - S_o) \tag{11}$$

where: H -enthalpy[kJ /kg];H_o-specific enthalpy at standard condition[kJ /kg];S- entropy [kJ /kgK];S_o- specific entropy at standard condition[kJ /kgK];T_o- ambient standard temperature [K];

The physical exergy of an ideal gas with constant specific heat c_p and specific heat ratio k can be written as:

$$ex_{ph} = c_p T_o \left[\frac{T}{T_o} - 1 - \ln\left(\frac{T}{T_o}\right) + \ln\left(\frac{P}{P_o}\right)^{\frac{k-1}{k}} \right] \tag{12}$$

where: P -pressure (atm), P_o – standard pressure (atm).

Chemical exergy: The chemical exergy is associated with the released of chemical composition of a system from that of the environment. Chemical exergy is equal to the maximum amount of work obtainable when the substance under consideration is brought from the environmental state (T_o, P_o) to the dead state (T_o, P_o, C_{0_i}) by processes involving heat transfer and exchange of substances only with the environment. The specific chemical exergy at P_o can be calculated by bringing the pure component in chemical equilibrium with the environment.
The chemical exergy can be calculated from [11, 12] as:

$$ex_{ch} = \sum x_n e_{ch}^n + RT_o \sum x_n \ln x_n \tag{13}$$

where: x_n - molar fraction of component n, e_{ch}^n – standard chemical exergy [kJ /kg]; R - universal gas constant [kJ /$kmolK$] .

The chemical exergies of gaseous fuels are computed from the stoichiometric combustion chemical reactions. The standard chemical exergies of various fuels can found in the literature.

Mass flow rates of the products and the reactants: Depending on the power output ($\dot{m}_{elect}$), and a fuel cell voltage (V_{cell}), and the stoichiometry of air(), the mass flow rates of the reactants and the products in the fuel cell can be easily evaluated from the equations used by Larminie and Dicks.

To calculate the mass flow rate of reactant air, we must to know the oxygen usage firstly. From the basic operation of the fuel cell, we know that four electrons are transferred for each mole of oxygen. So oxygen usage can be evaluated through the following equation:

$$Oxygen\ usage = \dot{m}_{O_2} \rightarrow$$

$$\dot{m}_{O_2} = \frac{32 \cdot 10^{-3} \cdot \dot{W}_{elect}}{4 \cdot F \cdot V_{cell}} = 8.29 \times 10^{-8} \left(\frac{\dot{W}_{elect}}{V_{cell}} \right) \tag{14}$$

At standard atmospheric conditions, the air molar analysis (%) would be: 77.48 N_2, 20.59 O_2, 0.03 CO_2 and 1.9 H_2O. Therefore molar proportion of air that is oxygen is approximately 0.21, and the molar mass of air is $28.97 \cdot 10^{-3}$ kg mol^{-1}. So the air inlet flow rate or mass flow rate of reactant air can be evaluated through the following equation:

$$Air\ inlet\ flow\ rate = \dot{m}_{air,R} \rightarrow$$

$$\dot{m}_{air,R} = \frac{28.97 \cdot 10^{-3} \cdot \lambda \cdot \dot{W}_{elect}}{0.21 \cdot 4 \cdot F \cdot V_{cell}} = 3.57 \times 10^{-7} \left(\frac{\lambda \dot{W}_{elect}}{V} \right) \tag{15}$$

The exit air flow rate or mass flow rate of the product air can be defined as the difference between the amount of air inlet flow rate and amount of oxygen usage:

Using equations (14), (15) this becomes:

$$Exit\ air\ flow\ rate = m_{air,P} \rightarrow$$

$$m_{air,P} = 3.57 \times 10^{-7} \left(\frac{\lambda \dot{W}_{elect}}{V} \right) - 8.29 \times 10^{-8} \left(\frac{\dot{W}_{elect}}{V} \right) \tag{16}$$

The hydrogen usage or mass flow rate of reactant hydrogen is derived in a way similar to oxygen, except that there are two electrons from each mole of hydrogen. So hydrogen usage can be evaluated through the following equation:

$$Hydrogen\ usage = \dot{m}_{H_2,R} \rightarrow$$

$$\dot{m}_{H_2,R} = \frac{2.02 \cdot 10^{-3} \cdot \dot{W}_{elect}}{2 \cdot F \cdot V_{cell}} = 1.05 \cdot 10^{-8} \cdot \left(\frac{\dot{W}_{elect}}{V} \right) \tag{17}$$

In a hydrogen-fed fuel cell, water is produced at the rate of one mole for every two electrons. The molecular mass of water is $18.02 \cdot 10^{-3}$ kg mole^{-1}. The amount of water produced by the fuel cell can be calculated by the following equation:

$$Water\ production = \dot{m}_{H_2O,P} \rightarrow$$

$$\dot{m}_{H_2O,P} = \frac{18.02 \cdot 10^{-3} \cdot \dot{W}_{elect}}{2 \cdot F \cdot V_{cell}} = 9.34 \cdot 10^{-8} \cdot \left(\frac{\dot{W}_{elect}}{V} \right) \tag{18}$$

Negligible to potential and kinetic energy effects on the fuel cell electrochemical process, the total exergy transfer per unit mass of each reactant and product consists of the combination of both physical and chemical exergies:

$$ex = ex_{ph} + ex_{ch} \qquad\qquad (19)$$

3. Discussion and results

The exergy analysis of a PEM fuel cell system is defined on the 1.2kW Nexa™ PEM power module taken from Ballard Power Systems Inc, the Nexa™ module installed at Fuel Cell Laboratory, Institute of Physics and Technology, Mongolian Academy of Sciences in 2010.

Nexa™ power module description: This module is capable of providing 1.2kW of unregulated DC output. The output module voltage level can vary from 43V at no load to about 26V at the full load. By the way an increasing of load, we were taken the increasing data of the current density at real time. The designed operating temperature in the stack is around 65°C at the full load. There are totally 47 cells connected in series in the stack. A individual fuel cell element consists of two electrodes, the anode and the cathode, separated by a polymer membrane electrolyte. Each of the electrodes is coated on one side with a thin platinum catalyst layer. The electrodes, catalyst and membrane together form the membrane electrode assembly.

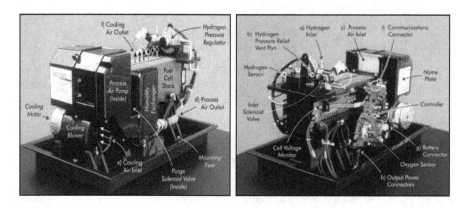

Figure 1. Main components and principle for operation of the Nexa™ PEM power module (Adapted from Ballard, Power systems inc., 2004).

A single fuel cell element produces about 1V at open-circuit and about 0.6V at full current output. The geometric area of the cells is 120 cm². The fuel is 99.99% hydrogen with no humidification, and the hydrogen pressure to the stack is normally maintained at 0.3 bar. Oxygen comes from the ambient air. The pressure of the oxidant air is 0.1 bar, and the air is humidified through a built in humidity exchanger to maintain membrane saturation and prolong the life of the membrane. Any drying of the PEM will greatly reduce the life of the fuel cell system. A humidity exchanger transfers both fuel cell product water and heat from the wet cathode outlet to the dry incoming air. The excess product water is discharged from the system, as both liquid and vapor in the exhaust. There is a small compressor supplying excess oxidant air to the fuel cell, and the speed of the compressor can be adjusted to match the power demand from the fuel cell stack.

The Nexa™ fuel cell module stack is air-cooled; the cooling fan draws air from the ambient surroundings in order to cool the fuel cell stack and regulate the operating temperature. Onboard sensors monitor system performance and the control board and microprocessor fully automate operation. The Nexa™ system also incorporates operational safety systems for indoor operation [13]. In figure 1 illustrates all components and subsystems of Nexa™ power module.

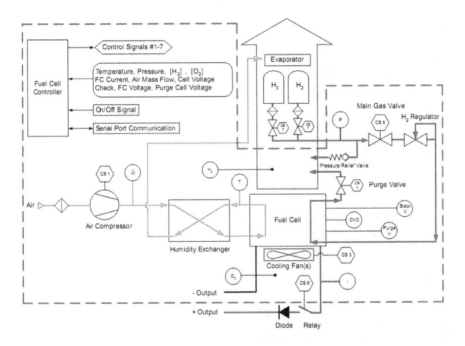

Figure 2. Schematic of the Nexa™ power module module (Adapted from Ballard, Power systems inc., 2004).

Figure 2 illustrates the schematic diagram of Nexa™ system. Hydrogen, oxidant air, and cooling air must be supplied to the unit, as shown in Figure 2. Exhaust air, product water and coolant exhaust is emitted.

The fuel-supply system, as shown in Figure 2, monitors and regulates the supply of hydrogen to the fuel cell stack. The fuel cell stack is pressurized with hydrogen during operation. The regulator assembly continually replenishes hydrogen, which is consumed in the fuel cell reaction. Nitrogen and product water in the air stream slowly migrates across the fuel cell membranes and gradually accumulates in the hydrogen stream. The accumulation of nitrogen and water in the anode results in the steady decrease in performance of certain key fuel cells, which are termed "purge cells". In response to the purge cell voltage, a hydrogen purge valve at the stack outlet is periodically opened to flush out inert constituents in the anode and restore performance. Only a small amount of hydrogen purges from the system, less than one percent of the overall fuel consumption rate. Purged hydrogen is discharged into the cooling air stream before it leaves the Nexa™ system, as shown in Figure 2. Hydrogen quickly diffuses into the cooling air stream and is diluted to levels many times less than the lower flammability limit. The hydrogen leak detector, situated in the cooling air exhaust, ensures that flammable limits are not reached.

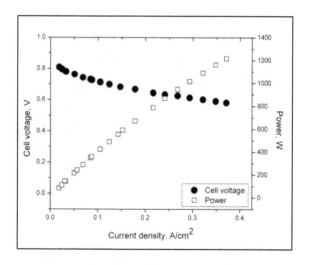

Figure 3. Polarization, power density curves at different temperatures and with different gas flows.

At high current levels, more heat is generated. It is important to keep the fuel cell stack temperature at a constant operating temperature; therefore, the fuel cell stack temperature has to be controlled. Fuel cell systems are either liquid-cooled or air-cooled. The Nexa™ fuel cell stack is air-cooled. A cooling fan located at the base of the unit blows air through vertical cooling channels in the fuel cell stack. The fuel cell operating temperature is maintained at 65°C by varying the speed of the cooling fan. The fuel cell stack temperature is measured at the cathode air exhaust, as shown in Figure 2. Hot air from the cooling system may be used for thermal integration purposes. Heat rejected in the air can be used for integration with metal hydrides, for evolving hydrogen. Hot air may also be used for space heating in some cases. The cooling system is also used to dilute hydrogen that is purposely purged from the Nexa™ module during normal operation.

Nexa™ system operation is automated by an electronic control system. The control board receives various input signals from onboard sensors. Input signals to the control board include: fuel cell stack temperature, hydrogen pressure, hydrogen leak concentrations, fuel cell stack current, air mass flow, fuel cell stack voltage and purge cell voltage.

Exergy analysis of NEXA™fuel cell module: The analysis is conducted on cell operating voltages from 0.001 to 0.84 V at air stoichiometrics from 13.0 to 2.1 in order to determine their effects on the efficiency of the fuel cell. The calculations of the physical and chemical exergies, mass flow rates and exergetic efficiency are performed at *temperature ratios* (T/T_0) (*Exit air flow rate = Air inlet flow rate - oxygen usage*) and *pressure ratios* (P/P_0) ranging from 0.93 to 1.13 and 7.44 to 4.91, respectively.

In Figure 3 shown calculated load characteristics represent cell voltage depending on current density (I-V curve).

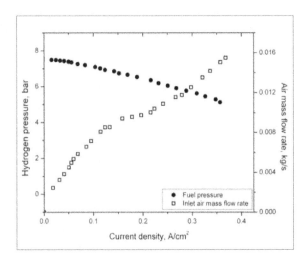

Figure 4. Reactants flow rates and hydrogen pressure of Nexa™ fuel cell module.

From the measured data, we calculated the cell voltage and current density by Eq. (19), Eq. (20).

$$V_{cell} = \frac{V}{47} \tag{20}$$

$$J = \frac{I}{120} \tag{21}$$

where: V – *output voltage* [V] ; *47 – number of stack,* J – *current density* [A/cm^2]; I – *output current* [A]; *120 – geometric area of the cell* [cm^2].

In Figure 4 illustrated the dependence of hydrogen pressure and mass flow rates of the inlet air. Mass flow rates of the inlet air and hydrogen were calculated from Eq. (15, 17), respectively. Hydrogen pressure data and air stoichiometric ratio values we taken from measured data. With increasing current density the hydrogen pressure decreases and its inlet air mass flow rate increases as shown in Figure 4.

Variation of stoichiometric air ratio (λ) illustrated in Figure 5a. The mass flow rate of the product air, can be defined as the difference between the amount of oxygen in the electrochemical reaction and the amount of oxygen consumed by reacting with hydrogen to produce water. The products, water flow rate and unused air calculated through Eq. (16, 18), respectively.

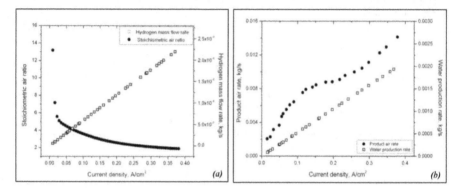

Figure 5. (a) Hydrogen mass flow rate and stoichiometric air ratio (b) Products flow rates of Nexa™ fuel cell module.

In Figure 5b illustrated, water production rate has small amount and production air rate increases depending on current density increase. Values of the chemical exergies for both the reactants and products are taken from published literature [10] and presented in Table 1.

	Chemical exergy, ex_{ch} (kJ/kg)				
Reactant/Product	Reactant Air	Reactant H_2	Product H_2O	Product Air	Reactant O_2
Nexa™ module	0	159138	2.5	8.58	-
MEA by Pt/MWCNT catalyst	-	159138	2.5	0	246

Table 1. Chemical exergy of the reactants and products of Nexa™ module and MEA by Pt/MWCNT catalyst.

The physical exergies of product water and product air calculated from Eq. (11, 19) was used to determine the physical exergies of inlet air and hydrogen. The values of fuel pressure and operation temperature taken from measured data.

The total exergy of the reactants and the products can be determined Eq. (15-18).

Finally, exergy efficiency was calculated by Eq. (2). The energy efficiency of the system can be calculated from Eq. (17) using the experimental $\dot{W}_{net}$ and $\dot{m}_{H_2,R}$ data [7]:

$$\eta_{energy,system} = \frac{\dot{W}_{elect}}{(HHV_{H_2} \cdot \dot{m}_{H_2})_R} \tag{22}$$

where: *HHV – higher heating value* [MJ/kg]; $\dot{m}_{H_2,R}$ – *reactant hydrogen mass flow rate which calculated by Equation (17)* .

Figure 6 illustrated the power productions and calculated energy, exergy efficiencies at different current density values. It can be seen that energy and exergy efficiencies decreases while power production increases.

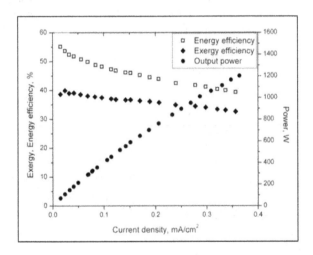

Figure 6. Energy, exergy efficiency and power of Nexa™ fuel cell module depending on current density.

Energy and Exergy efficiency of 1.2 kW Nexa™ power module decreases depending on current density increases. Energy efficiencies vary between 41 – 55%, exergy efficiency from 33% to 42% at the current density of 0.02 - 0.36A/cm² respectively. From the Figure 6, increasing of flow rates and decreasing of hydrogen pressure caused to the decreasing of the energy and exergy efficiencies of the module.

Fabrication of Membrane-electrode assembly by the synthesized Pt/MWCNT catalyst: The preparation of the MEA was carried out using a Nafion® 117 membrane from Dupont. For both electrodes an ink solution was prepared using a method slightly modified from the one reported by Gottesfeld et al., [14]. MEAs with an active electrode area of 25 cm² were fabricated by airbrushing the catalyst ink onto one side of the Nafion membrane, heated to and kept at 120 °C. For the cathode, 100 mg of the catalyst were dispersed in 0.5 ml ultrapure water, isopropanol and 1 ml Nafion® 5 % solution under sonication. The dispersion was stirred with a high shear mixer at 7000 rpm. For the anode, 200 mg of Pt on carbon (20 wt% Pt, Alfa Aesar HISPEC® 3000) were dispersed in 4 ml of H_2O and 2 ml isopropanol. 1.2 ml of Nafion® 5 % solution was added, and the solution was dispersed by sonication and stirred with a high shear mixer. The inks were filled into an airbrush pistol (Evolution by Harder & Steenbeck) and sprayed successively onto the heated membrane surface, allowing each layer to dry for 10 seconds. Hydrogen and oxygen or air reactants are fed to the anode and cathode compartments, respectively, with or without pre - humidifying. Usually, the cell is conditioned by operating at low loading to activate the MEA. After that, the polarization curve is recorded galvanostatically by stepping the current from zero to the maximum test current density (Figure 7). The polarization curve is effective and intuitional to characterize the performance. However, separation of the electrochemical and ohmic contributions to polarization requires additional experimental techniques. This can be done by measuring the electrochemical impedance spectroscopy. The flow rates of both gases were adjusted to H_2/O_2 55/25 sccm and 83/38 sccm, respectively, and the cell temperatures varied between 25°C, 50°C, and 65°C. Hydrogen was loaded with water in a humidifier (25 °C) and fed into the anode. The voltage at each current density is allowed to stabilize before measurement. MEAs were conditioned overnight until a steady state current achieved at a potential of 0.6V. The operation of the fuel cell test station was controlled and monitored by LabView programs [15].

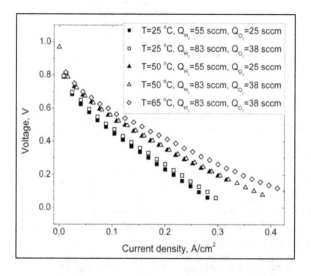

Figure 7. Polarization and power density curves at different temperatures and with different gas flows.

Exergy analysis on the Membrane-electrode assembly by the synthesized Pt/MWCNT catalyst: Based on this exergy analysis of Nexa™ power module, we calculated the exergy efficiency on membrane-electrode assembly by the synthesized Pt/MWCNT catalyst. The MEAs fabricated from the Pt/MWCNT performed well, and the polarization curves and power densities in different operation conditions can be found in Figure 7.

At the current density up to 0.2 A/cm², the exergy efficiency decreases from 72% to 35% as shown in Figure 8. We can explain that the exergy efficiency of 72% at the low current density is caused by the only one MEA and to feed pure oxygen (O_2). Therefore, the activation loss is main effect to the rapid decrease of the exergy efficiency.

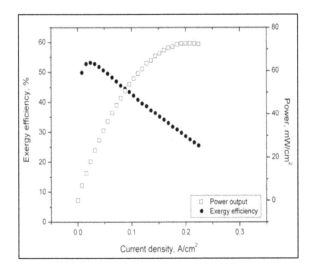

Figure 8. Exergy efficiency and power output of PEM fuel cell.

4. Conclusion

We performed the exergy analysis of 1.2 kW Nexa™ power module at variable operating conditions such as a different temperatures, pressures, cell voltages and stoichiometry.

The total exergy of the reactants and the products consist of both physical and chemical exergies, which are calculated for each element in the electrochemical process. Through the experimental data, we were calculated flow rates, energy efficiency, physical, chemical exergy and exergy efficiency.

The results provided that exergy efficiencies of the PEM fuel cell module less than energy efficiencies. From our calculation, it is recommended that the PEM fuel cell should operate at stoichiometric ratios less than 4 in order to optimize the relative humidity level in the product air and to avoid the membrane drying out at high operating temperatures. Exergy efficiency of the NEXA™ PEM fuel cell can be improved through increasing the fuel cell operating temperature (in our case up to 72°C), and improved by a higher operating pressure. From the result of the exergy efficiency of membrane - electrode assembly of the Pt/ MWCNT higher than Nexa™ power module, it might be explained fed pure oxygen (O_2), in other hand a high air stoichiometry could be improve a fuel cell exergitic efficiency.

Author details

G. Sevjidsuren, E. Uyanga, B. Bumaa, E. Temujin, P. Altantsog and D. Sangaa

*Address all correspondence to: sevjidsuren@ipt.ac.mn

Department of Material Science and Nanotechnology, Institute of Physics and Technology, Mongolian Academy of Sciences, Mongolia

References

[1] Barbir, F. (2005). *PEM Fuel Cells Theory and Practice*, Elsevier Academic Press.

[2] Basu, S. (2007). *Recent Trends in Fuel Cell Science and Technology*, Anamaya Publishers, New Delhi, India.

[3] Gencoglu, M. T., & Ural, Z. (2009). Design of PEM fuel system for residential application. *J. Hydrogen Energy*, 34, 5242-5248.

[4] Zhang, Jiujun. (2008). *PEM fuel cell electrocatalysts and catalyst layers*, Springer, Verlag London Ltd.

[5] Smith, J. M., Van Ness, H. C., & Abbott, M. M. (2001). *Introduction to chemical engineering thermodynamics*, McGraw-Hill.

[6] Dincer, I., & Rosen, M. A. (2007). *Exergy: Energy, environment and sustainable development*.

[7] Zang, Z., Huang, X., & Jiang, J. (2006). An improved dynamic model considering effects of temperature and equivalent internal resistance for PEM fuel cell power modules. *J. Power Sources*, 161, 1062-1068.

[8] Kazim, Ayoub. (2004). Exergy analysis of a PEM fuel cell at variable operating conditions. *Energy Conversion and Management*, 45, 1949-1961.

[9] Kirubakaran, A., Sh, Jain, & Nema, R. K. (2009). The PEM Fuel Cell System with DC/DC Boost Converter: Design, Modeling and Simulation. *J. Recent Trends in Engineering*, 1, 157-161.

[10] Larminie, J., & Dicks, A. (2003). *Fuel cell systems explained*, John Wiley & Sons Ltd.

[11] Cengel, Y., & Boles, M. (1994). *Thermodynamics-an engineering approach* (2nd ed.), Mc Graw-Hill, Inc.

[12] Smith, M., Van Ness, H. C., & Abbott, M. M. (2001). *Introduction to chemical engineering thermodynamics*, McGraw-Hill Inc., New York.

[13] Nexa™ . (2003). *(310-0027) Power Module User's Manual*, Ballard Power Systems Inc.

[14] Wilson, M., & Gottesfeld, S. (1992). Thin film catalyst layers for polymer electrolyte fuel cell electrodes. *J Appl Electrochem*, 22, 1-7.

[15] Sevjidsuren, G. (2011). *Dissertation, Study on Processing of Proton Exchange Membrane Fuel Cell*, Ulaanbaatar, Mongolia.

Chapter 3

Thermodynamics Assessment of the Multi-Generation Energy Production Systems

Murat Ozturk

Additional information is available at the end of the chapter

1. Introduction

The efficiency of the solar multi-generation energy production system has great significance due to the limited supply of available energy from solar radiation as well as impact on system production performance, operation cost and environmental concerns. Thus, a good understanding of the efficiency of the whole system and its components is necessary for the multi-functional system installation. In this regard, the First Law of Thermodynamics based efficiency also known as energy efficiency may lead to inadequate and also misleading consequences, since all energy transfers are taken to be equal and the ambient temperature is not taken into consideration. The Second Law of Thermodynamics defines the energy conversation limits of this available energy based on irregularities between different forms of energies. The quality of the available energy is highly connected with the reference environment, which is often modeled as the ambient environment, as well as the success level of this conversion capacity; and needs to be considered to prevent any incomplete and/or incorrect energy conversation results. Quality of the energy should be given as an examining the work potentials of the initial and final stages of an investigated system. Such analysis is called as exergy analysis, which gives the amount of an energy that may be totally converted into useful work. Exergy (also called as an available energy or availability) of an investigated system is defined using the thermodynamics principles as the maximum amount of work which can be produced by a system or a flow of matter or energy as it comes to equilibrium with a reference environment [1-3]. It is well known that one of the important uses of the exergy analysis in engineering processes is to determine the best theoretical performance of the system.

The useful work potential of the system is reduced by the irreversibilities and the corresponding amount of energy becomes unusable [4]. The entropy generation give the effects of these irreversibilties in the investigation system during a process and helps compare the each component in the system based on how much they contribute to the operation ineffi- ciencies of the whole system. Thus, entropy generations of the system components needs to be evaluated to determine the whole system efficiency. Even though energy analysis of the system is the most commonly used method for examining energy conversation systems, its only concerned with the conservation of energy, which neither takes the corresponding en- vironmental conditions into account, nor provides how, where and why the system per- formance degrades for the operated system. Also, the energy analysis of the system only measures the quantity of energy and does not reveal the full efficiencies of the process [5]. Thus, in this scientific study, the multi-generation system is examined with exergy analysis in order to give the true efficiency of the whole system and its components by determining the irreversibilities in the each process, and how nearly the respective performance ap- proach ideal conditions. By using the energy and exergy analysis, magnitude of the losses, and their causes and locations are identified by investigating the sites of exergy destruction in order to make improvements to the whole system and its components [6].

2. System Description

The whole system and its components are given for the solar multi-generation energy pro- duction system. This system can be divided into four subsystems; i-) parabolic trough collec- tor, ii-) organic Rankine cycle (ORC), iii-) electrolyzer and iv-) absorption cooling and heating. The schematic diagram of the multi-generation system is given in the Figure 1. The main outputs of the given system are electricity, hydrogen, oxygen, heating, cooling and hot water. Thermal energy of the solar radiation is collected and concentrated using a parabolic trough collector in order to produce electricity, heating-cooling and hot water from ORC, absorption system and hot water collection tank, respectively. Another important purpose of this solar multi-generation system is producing of hydrogen. Stored hydrogen can be used in a PEM fuel cell to produce power in the night time. Thus, electricity can be pro- duced continuously for 24 hours. A part of produced electricity from the organic Rankine cycle is used to run the PEM (proton exchange membrane) electrolysis system, which re- quires heat at nearly 80 °C and electricity as an input. The thermal energy is used in the PEM electrolysis to decrease the electricity demand of the electrolysis system. Heat require- ments of the PEM electrolyzer system are supplied from generator waste heat. The hydro- gen separator separates hydrogen from the steam by using hydrogen separation membrane. The produced hydrogen stream is then cooled to 40°C with the help of the cooling water. The produced hydrogen is compressed in a four-stage compressor, through intercooling to 40°C. The product gases (which is 99.9 wt% H_2) exit from the store tank at 506.5 kPa and 85°C. When the weather conditions are not favorable or additional power is needed, the stored hydrogen can be used in order to generate power. In addition, the outputted oxygen from the high temperature electrolysis is stored in a separate storage tank. The produced oxygen from the high temperature electrolysis system is cooled to 45°C via the cooling wa- ter. Similarly, the cooling water used in the oxygen cooler has an exit temperature of 80°C,

The by-product oxygen has relatively negligible energy and exergy content, and can be used for the other purposes or can be sold.

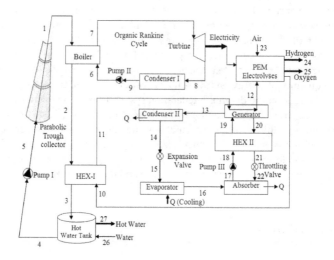

Figure 1. The Solar multi-generation energy production system

3. Thermodynamic Analysis

In this research, the general mass, energy and exergy balance equations to find the energy and exergy inputs and outputs, the rate of exergy decrease, the rate of irreversibility and the energy and exergy efficiencies for the solar multi-generation energy production system are given. In general, thermodynamic balance equation for a quantity in a process can be given as

$$Input + Generation - Output - Consumption = Accumulation \tag{1}$$

where input and output gives to quantities entering and exiting through the system boundary, respectively, generation and consumption gives to quantities produced or consumed within the system, respectively and accumulation gives to potential build-up of the quantity within the system [7]. The general mass balance equation can be given in the rate form.

$$\sum \dot{m}_{in} = \sum \dot{m}_{out} \tag{2}$$

where $\dot{m}$ is the mass flow rate, and the subscripts in and out shows inlet and outlet flows, respectively. Assuming the absence kinetic, potential and chemical exergy terms, the general energy balance for the multi-generation system is formulated as follows;

$$\dot{Q} + \sum_i \dot{m}_{in,i}\left(h_i + \frac{v_i^2}{2} + gz_i\right) = \sum_e \dot{m}_{out,e}\left(h_e + \frac{v_e^2}{2} + gz_e\right) + \dot{W} \tag{3}$$

where $\dot{Q}$ and $\dot{W}$ represents the heat and work rates, respectively, and h is the specific enthalpy at the chosen state. Considering a system at rest relative to the environment, kinetic and potential terms can be ignored,

$$\dot{Q} + \sum_i \dot{m}_{in,i}h_i = \sum_e \dot{m}_{out,e}h_e + \dot{W} \tag{4}$$

The entropy balance can also be expressed on a time rate basis as

$$\dot{S}_{in} + \dot{S}_{gen} = \dot{S}_{out} \tag{5}$$

$$\dot{S}_{gen} = \dot{m}\Delta S \tag{6}$$

where $\dot{S}$ is the entropy flow or generation rate. The amount transferred out of the boundary must exceed the rate in which entropy enters, the difference being the rate of entropy generation within the boundary due to associated irreversibilities.

The system components irreversibility and also recommended ways to improve the efficiencies of them can be evaluated by using exergy analysis. The exergy balance of the multi-generation system components is given as follows;

$$\sum \dot{Ex}_{in} = \sum \dot{Ex}_{out} + \dot{Ex}_D \tag{7}$$

$$\sum_i \dot{m}_i ex_i + \dot{Ex}_Q = \sum_e \dot{m}_e ex_e + \dot{Ex}_W + \dot{Ex}_D \tag{8}$$

where subscripts i and e are the specific exergy of the control volume inlet and outlet flow, $\dot{Ex}$ is the exergy rate, $\dot{Ex}_Q$ and $\dot{Ex}_W$ are the exergy flow rate associated with heat transfer and work, ex is the specific flow exergy of the process and $\dot{Ex}_D$ is the exergy destruction rate.

$$\dot{Ex}_Q = \left(1 - \frac{T_o}{T_i}\right)\dot{Q}_i \tag{9}$$

$$\dot{Ex}_W = \dot{W} \tag{10}$$

$$ex = ex_{ke} + ex_{pe} + ex_{ph} + ex_{ch} \tag{11}$$

where ex_{ke} is the kinetic exergy, ex_{pe} is the potential exergy, ex_{ph} is the physical exergy and ex_{ch} is the chemical exergy. Since the variation of the kinetic, potential and chemical exergy is considered negligible in this study. The physical exergy

$$ex_{ph,i} = (h_i - h_o) - T_o(s_i - s_o)$$ (12)

The exergy rate of a material flow is given as follows.

$$\dot{E}x_i = \dot{m}ex_i$$ (13)

The difference being the rate of exergy destruction (or lost work) within the boundary due to associated irreverisibilities which can be calculated based on Gouy-Stodola theorem. The exergy destruction in the component i should be given as follows;

$$\dot{E}x_{D,i} = T_o \Delta S_{i,net}$$ (14)

where $\Delta S_{i,net}$ is the specific entropy change for the process. The exergy loss ratio of the system components is given as follows to compare of these components by using exergy analysis view point.

$$\dot{E}x_{LR} = \frac{\dot{E}x_{D,com}}{\dot{E}x_{D,sys}}$$ (15)

where $\dot{E}x_{D,com}$ is exergy destruction of the system components and $\dot{E}x_{D,sys}$ is the exergy destruction of the overall system. The exergy destruction rate for the each component and overall of the multi-generation system is given in the Table 1 according to given above procedure.

Components of the system	Expression of exergy destruction rate
Parabolic trough collector	$\dot{E}x_{D,PTC-I} = \dot{E}x_5 - \dot{E}x_1 + \dot{E}x_{Solar}^Q$
Boiler	$\dot{E}x_{D,Boil-I} = \dot{E}x_1 + \dot{E}x_6 - \dot{E}x_2 - \dot{E}x_7$
HEX I	$\dot{E}x_{D,HEX-I} = \dot{E}x_2 + \dot{E}x_{10} - \dot{E}x_3 - \dot{E}x_{11}$
Hot Water Tank	$\dot{E}x_{D,HWT} = \dot{E}x_3 + \dot{E}x_{26} - \dot{E}x_4 - \dot{E}x_{27}$
Pump I	$\dot{E}x_{D,P-I} = \dot{E}x_4 - \dot{E}x_5 + \dot{W}_{P-I}$
Turbine	$\dot{E}x_{D,HPT} = \dot{E}x_7 - \dot{E}x_8 - \dot{W}_{HPT}$
Condenser I	$\dot{E}x_{D,Con-I} = \dot{E}x_8 - \dot{E}x_9 - \dot{E}x_{Con-I}^Q$
Pump II	$\dot{E}x_{D,P-I} = \dot{E}x_9 - \dot{E}x_6 + \dot{W}_{P-II}$
PEM Electrolyses system	$\dot{E}x_{D,PEM} = \dot{E}x_{12} + \dot{E}x_{23} - \dot{E}x_{PEM}^Q + \dot{W}_{PEM} - \dot{E}x_{24,H_2} - \dot{E}x_{25,O_2}$
Generator	$\dot{E}x_{D,Gen} = \dot{E}x_{11} + \dot{E}x_{19} - \dot{E}x_{13} - \dot{E}x_{12} - \dot{E}x_{20}$
Condenser II	$\dot{E}x_{D,Con-II} = \dot{E}x_{13} - \dot{E}x_{14} - \dot{E}x_{Con-II}^Q$
Expansion valve	$\dot{E}x_{D,ExV} = \dot{E}x_{14} - \dot{E}x_{15}$
Evaporator	$\dot{E}x_{D,Ev} = \dot{E}x_{15} - \dot{E}x_{16} + \dot{E}x_{Ev}^Q$
Absorber	$\dot{E}x_{D,Ab} = \dot{E}x_{16} + \dot{E}x_{22} - \dot{E}x_{17} - \dot{E}x_{Ab}^Q$
Pump III	$\dot{E}x_{D,P-III} = \dot{E}x_{17} - \dot{E}x_{18} + \dot{W}_{P-III}$
Throttling valve	$\dot{E}x_{D,Thv} = \dot{E}x_{21} - \dot{E}x_{22}$
HEX II	$\dot{E}x_{D,HEX-II} = \dot{E}x_{18} + \dot{E}x_{20} - \dot{E}x_{19} - \dot{E}x_{21}$

Table 1. Exergy destruction rates for the multi-generation energy production system

3.1. Energy efficiency

The energy efficiency of the process is defined as the ratio of useful energy produced by the process to the total energy input. In this paper, energy efficiencies for five different systems are considered: parabolic trough collector, organic Rankine cycle, hydrogen production, absorption cooling and heating sub-system, overall multi-generation system as shown below

$$\eta_{PTC} = \frac{\dot{Q}_1 + \dot{Q}_5}{\dot{Q}_{solar}} \tag{16}$$

$$\eta_{org-Rankine} = \frac{\dot{W}_{net,org-Rankine}}{\dot{Q}_{boiler}} \tag{17}$$

$$\eta_{hydrogen} = \frac{\dot{m}_{H_2} LHV\, H_2}{\dot{Q}_{Gen} + \dot{W}_{Turbine}} \tag{18}$$

$$\eta_{absorption} = \frac{\dot{Q}_{cooling} + \dot{Q}_{heating}}{\dot{Q}_{HEX-I} + \dot{W}_{P-III}} \tag{19}$$

$$\eta_{system} = \frac{\dot{W}_{org-Rankine} + \dot{m}_{H_2} LHV\, H_2 + \dot{Q}_{cooling} + \dot{Q}_{heating} + \dot{Q}_{hotwater}}{\dot{Q}_{PTC}} \tag{20}$$

A coefficient of performance (COP) term can be used to expressing of the energetic performance of the absorption sub-system.

$$COP = \frac{\dot{Q}_{cooling}}{\dot{W}_{pump-III} + \dot{Q}_{gen}} \tag{21}$$

where $\dot{W}_p$ is the pumping power requirement, and it is usually neglected in the COP calculations and $\dot{Q}_{gen}$ is the rate of heat inputted to the generator.

3.2. Exergy Efficiency

The exergy efficiency of the process is the produced exergy from the system output that is divided by the exergy system input and it can also be expressed by the aforementioned subsystems as follows;

$$\psi_{PTC} = \frac{\dot{E}x_1^Q + \dot{E}x_2^Q}{\dot{E}x_{solar}^Q} \tag{22}$$

$$\psi_{org-Rankine} = \frac{\dot{W}_{org-Rankine}}{\dot{E}x_{boiler}^Q} \tag{23}$$

$$\psi_{hydrogen} = \frac{\dot{E}x_{H2}}{\dot{E}x_{Gen}^Q + \dot{W}_{Turbine}} \tag{24}$$

$$\psi_{absorption} = \frac{\dot{E}x^Q_{cooling} + \dot{E}x^Q_{heating}}{\dot{E}x^Q_{HEX-I} + \dot{W}_{P-III}} \tag{25}$$

$$\psi_{system} = \frac{\dot{W}_{org,Rankine} + \dot{E}x_{H2} + \dot{E}x^Q_{cooling} + \dot{E}x^Q_{heating} + \dot{E}x^Q_{hotwater}}{\dot{E}x^Q_{PTC}} \tag{26}$$

The exergy efficiency equations for the solar-based multi-generation energy production system components are given in the Table 2. The exergetic performance of the absorption subsystem should be formulated in forms of the exergetic COP.

$$COP_{ex} = \frac{\dot{E}x^Q_{cooling}}{\dot{W}_{pump-III} + \dot{E}x^Q_{gen}} \tag{27}$$

Components of the system	Exergy efficiency
Parabolic trough collector	$\psi_{PDC-i} = \dfrac{\dot{E}x_1 - \dot{E}x_5}{\dot{E}x^Q_{solar}}$
Boiler	$\psi_{HEX-I} = \dfrac{\dot{E}x_7 - \dot{E}x_6}{\dot{E}x_1 - \dot{E}x_2}$
HEX I	$\psi_{HEX-III} = \dfrac{\dot{E}x_{11} - \dot{E}x_{10}}{\dot{E}x_2 - \dot{E}x_3}$
Hot Water Tank	$\psi_{HWT} = \dfrac{\dot{E}x_{27} - \dot{E}x_{26}}{\dot{E}x_3 - \dot{E}x_4}$
Pump I	$\psi_{Pump-I} = \dfrac{\dot{E}x_5 - \dot{E}x_4}{\dot{W}_{Pump-I}}$
Turbine	$\psi_{HPT} = \dfrac{\dot{W}_{HPT}}{\dot{E}x_7 - \dot{E}x_8}$
Condenser I	$\psi_{Con-I} = \dfrac{\dot{E}x^Q_{Con-I}}{\dot{E}x_8 - \dot{E}x_9}$
Pump II	$\psi_{Pump-I} = \dfrac{\dot{E}x_6 - \dot{E}x_9}{\dot{W}_{Pump-II}}$
PEM Electrolyses system	$\psi_{HTSE} = \dfrac{\dot{m}_{H_2} LHV_{H_2}}{\dot{E}x_{12} + \dot{W}_{Electricity}}$
Generator	$\psi_{Gen} = \dfrac{\dot{E}x_{13} + \dot{E}x_{20} - \dot{E}x_{19}}{\dot{E}x_{11} - \dot{E}x_{12}}$
Condenser II	$\psi_{Con-II} = \dfrac{\dot{E}x^Q_{Con-III}}{\dot{E}x_{13} - \dot{E}x_{14}}$
Expansion valve	$\psi_{ExV} = \dfrac{\dot{E}x_{13}}{\dot{E}x_{12}}$
Evaporator	$\psi_{Eva} = \dfrac{\dot{E}x^Q_{col}}{\dot{E}x_{16} - \dot{E}x_{15}}$
Absorber	$\psi_{Ab} = \dfrac{\dot{E}x^Q_{Ab}}{\dot{E}x_{16} + \dot{E}x_{22} - \dot{E}x_{17}}$
Pump III	$\psi_{Pump-III} = \dfrac{\dot{E}x_{18} - \dot{E}x_{17}}{\dot{W}_{Pump-III}}$
Throttling valve	$\psi_{ThV} = \dfrac{\dot{E}x_{22}}{\dot{E}x_{21}}$
HEX II	$\psi_{HEX-II} = \dfrac{\dot{E}x_{19} - \dot{E}x_{18}}{\dot{E}x_{20} - \dot{E}x_{21}}$

Table 2. Exergy efficiency equations for the system components

4. Result and Discussion

A software code in EES (Engineering Equation Solver) [8] is created to analyze a baseline model with respect to the balance equations given in Table 2. The ambient conditions are assumed to be 25 °C and 100 kPa for the analysis. A widely used refrigerant R134a is used in the absorption cooling system and water in the other sub-systems. The assumptions for the analysis are given as follow.

• Steady-state conditions with no chemical or nuclear reactions are assumed for all components in the cycle.

• Heat loss and pump work as well as kinetic and potential energies are considered negligible.

• Expansion valve and throttling valve are assumed to be isenthalpic and the heat transfer and pressure drops in the tubes connecting the components are neglected since there are short.

The values for the exergy destruction rates (kW), exergy destruction ratio (%), exergy efficiency (%) and the power or heat transfer rate of the solar multi-generation energy production system are given in the Table 3. Exergy destruction rate indicates the reduction in energy availability; however, it cannot be used to investigate the energy and exergy utilization performance of the system processes. The exergy efficiencies of the system components are more useful for determining exergy losses.

	Exergy destruction rate (kW)	Exergy destruction ratio (%)	Exergy efficiency (%)	Power or heat transfer rate (kW)
Parabolic trough collector	1892	46.89	34.21	18798
Boiler	564.1	13.98	90.92	18798
HEX-I	337.6	8.37	82.89	3351
Hot water tank	310.2	7.69	28.19	2593
Pump-I	45.11	1.12	56.03	74.37
Turbine	259.3	6.43	93.6	3792
Condenser-I	81.67	2.02	25.71	12734
Pump-II	114.7	2.84	58.77	175.5
PEM electrolysis system	202.4	5.02	37.03	781.5
Generator	37.03	0.92	75.42	750
Condenser-II	19.91	0.49	21.94	604.87
Expansion valve	0.15	0.00	97.36	1.69

	Exergy destruction rate (kW)	Exergy destruction ratio (%)	Exergy efficiency (%)	Power or heat transfer rate (kW)
Evaporator	56.31	1.40	40.84	571.1
Absorber	98.25	2.43	21.42	696.2
Pump-III	8.24	0.20	34.41	12.92
Throttling valve	1.16	0.03	96.58	17.34
HEX-II	6.72	0.17	56.53	109.1

Table 3. Thermodynamic analysis data of the multi-generation energy production system devices

Based on the baseline analysis, the exergy efficiencies associated with the system and whole system are given in the Fig. 2. As seen in the Fig. 2, the solar parabolic trough collector and condenser are calculated to have the lowest exergy efficiency as 17 and 22%, respectively. This is associated with concentrating losses, high temperature differences and phase chance which results in more entropy generation between the inlet and outlet streams.

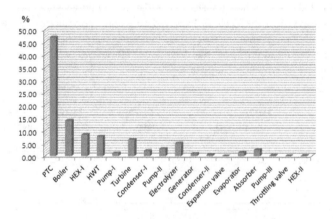

Figure 2. Exergy efficiencies of the multi-generation energy production system components

The COP and COPex of the single effect absorption refrigeration system are calculated as 0.7586 and 0.3321, respectively. The COPex is lower than COP, due to the considerable irreversibilities occurring in the absorption cycle. Energy and exergy efficiency results for the absorption system components are compared to the experimental studies [9-11] and a reasonably good agreement are found.

Parametric studies have also been conducted, by analyzing the changes in exergy efficiencies of the system components with respect to changes in the ambient temperature. The exergy efficiencies for the ambient temperature ranges of 10 °C to 30 °C can be seen in the Fig. 3 to 5 for the parabolic trough collector, organic Rankine cycle and absorption cooling-heating cycle, respectively.

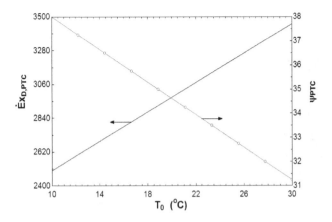

Figure 3. Exergy destruction rate and exergy efficiency of the parabolic trough collector (PTC) depending on ambient temperature changes

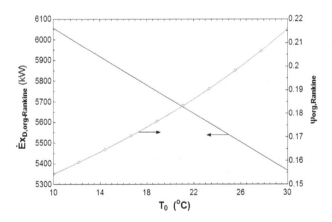

Figure 4. Exergy destruction rate and exergy efficiency of the organic Rankine cycle (ORC) depending on ambient temperature changes

Although ambient temperature increases, exergy destruction of the parabolic trough collector and absorption system are increases and exergy efficiencies of these components decreases. The variations of exergy destruction rate and exergy efficiency of these components according to the ambient temperature remain almost linear. These results are expected since the exergy destruction rate and exergy efficiency of the process are generally inversely proportional properties. The variations of the exergy destruction rate and also exergy efficiency in concern with the ambient temperature for the organic Rankine cycle are given in the Fig. 4.

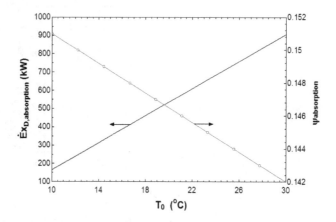

Figure 5. Exergy destruction rate and exergy efficiency of the absorption cooling and heating system depending on ambient temperature changes

5. Conclusions

In the present study, a solar multi-generation system for electricity, hydrogen, oxygen, heat water production and space heating and cooling is proposed and examined with respect to exergy analysis in order to determine the magnitude of losses, and their causes and locations by determining the irreversibility in each cycle and the whole system. In addition to that, exergy efficiency of the system components is evaluated to show how the system reaches a real operating condition. Some parametric studies are given in order to investigate the effects of varying operating conditions such as ambient temperature.

Author details

Murat Ozturk[*]

Address all correspondence to: muratozturk@sdu.edu.tr

Department of Physics, Faculty of Art and Sciences, Suleyman Demirel University, 32260, Isparta, Turkey

References

[1] Moran, M. J. (1982). Availability analysis: A guide to efficiency energy use. *Englewood Cliffs, NJ: Prentice-Hall.*

[2] Kotas, T. J. (1995). The Exergy Method of Thermal Plant Analysis. *Reprint ed. Krieger, Malabar, FL.*

[3] Szargut, J., Morris, D. R., & Steward, F. R. (1988). Exergy Analysis of Thermal, Chemical and Metallurgical Processes. Hemisphere, New York.

[4] Arcaklioglu, E., Çavuşoglu, A., & Erisen, A. (2005). An algorithmic approach towards finding better refrigerant substitutes of CFCs in terms of the second law of thermodynamics". *Energy Conversion and Management*, 46, 1595-1611.

[5] Yumrutas, R., Kunduz, M., & Kanoglu, M. (2002). Exergy analysis of vapor compression refrigeration systems. *Exergy, an International Journal*, 2, 266-272.

[6] Dincer, I., & Rosen, M. A. (2007). Exergy: Energy, Environment and Sustainable Development". Elsevier, Oxford, UK.

[7] Dincer, I., & Kanoglu, M. (2010). Refrigeration Systems and Applications. John Wiley and Sons, West Sussex, UK.

[8] Klein & S.A. Engineering Equation Solver (EES). *Academic Commercial, F-Chart Software, 2010 www.fChart.com.*

[9] Ataer, O. E., & Gogus, Y. (1991). Comparative study of irreversibilities in an aqua-ammonia absorption refrigeration system. *International Journal of Refrigeration*, 14, 86-92.

[10] Dincer, I., & Dost, S. (1996). A simple model for heat and mass transfer in absorption cooling systems (ACSs). *International Journal of Energy Research*, 20, 237-43.

[11] Dincer, I., & Kanoglu, M. (2010). Refrigeration Systems and Applications. Wiley Publication, United Kingdom.

Chapter 4

Multi-Level Mathematical Modeling of Solid Oxide Fuel Cells

Jakub Kupecki, Janusz Jewulski and
Jarosław Milewski

Additional information is available at the end of the chapter

1. Introduction

In recent years, numbers of questions concerning energy generation have arisen. Emission levels, delivery security, and diversification of the portfolio of technologies have been extensively discussed. Well-established generation based on fossil fuels in large-scale power stations is criticized for big environmental impacts, and limited sustainability due to high fraction of process losses. Not only emissions, but also extraction of resources, alternation of the landscape, transmission and distribution inefficiencies are often pointed as the main downside. As a solution for rapidly increasing energy consumption, and emerging threat of current resources depletion, distributed generation based on highly efficient micro- and small-system was proposed. Moreover, combined heat and power (CHP) units with high achievable efficiency are seen as possible substitutes for stand-alone electricity generators. Most of technologies from that group are currently under development, however selected systems are already reaching market availability. In 2004 European Commission indicated selected systems, with guidelines for promotion and development of highly efficient co-generative units [1]. List of technologies, which can provide high electrical and overall efficiency with limited environmental impacts, includes the following:

- Combined cycle gas turbine with heat recovery

- Steam backpressure turbines

- Steam condensing extraction turbines

- Gas turbines with heat recovery

- Internal combustion engines

- Microturbines

- Stirling engines

- Fuel cells

- Steam engines

- Organic Rankine cycles

- Any other type of technology of combination of thereof falling under the definition laid in the directive

Further studies were devoted to finding the optimal technology for micro- and small scale-systems suitable for CHP applications. It should be noted that, to distinguish between selected systems scales, terms micro and small were introduced. In the EU Combined Heat and Power directive [1] the earlier term refers to units with nominal power output 50 kW$_{el}$, while in literature it usually covers systems with nominal power output of single kW$_{el}$ [2,3,4]. The later usually refers to system with output of tens of kW$_{el}$.

It was found that three groups of technologies are especially interesting from the technical and economical point of view for systems with single kilowatts power output, namely:

- Internal combustion engines

- Stirling engines

- Fuel Cells (PEFC and SOFC)

While different energy generating systems with internal combustion and Stirling engines are a well-established technology, fuel cells in stationary generation have been known for not more than two decades. Even though technology is not yet mature, numerous demonstration systems have already been operated allowing to gain operating experience. The main reason to consider fuel cells as an alternative to other generation systems is high electrical performance due to the direct conversion of chemical energy of a fuel into electricity.

Evaluation of PEFC and SOFC for micro-CHP application was recently presented [3]. Authors underlined high efficiency and mulifuel capabilities of the SOFC. Additionally, in case of low-temperature cells, such as PEFC, partial internal reforming cannot be done, hence efficiency penalty due to external reforming is observed next to limited fuel flexibility. Moreover, SOFCs offer utilization of high temperature heat in co-generative systems [4]. Substantial part of the high-grade heat can be recovered from the anodic and cathodic gas streams leaving the SOFC stack at elevated temperature [5] for hot tap water supply or heating purposes [6]. Taking into account these advantages, SOFC technology has been selected for futher analysis with different modeling techniques.

2. Selected systems with Solid Oxide Fuel Cells

Over the years fuel cell technology proved to be feasible in a number of applications, including portable energy generation, transportation, stationary back-up systems and energy generators in space missions among others Currently, selected fuel cells such as SOFCs are considered as suitable conversion systems for the clean and sustainable energy generation. Development of such systems requires proper modeling approaches, construction of high fidelity numerical simulators and tools able to provide clear insight into various aspects of the system operation.

Selected units with SOFCs have already reached proof-of-the-concept stage of development, and in some cases units are already available in for sale. Comparison of market-available systems is presented in Table 1, based on available data [7,8,9,10,11,12,13]. It should be emphasized, that using fuel cells for electricity only generation in micro- and small-units is not economically feasible, therefore it was not considered. SOFC-based electricity-only generation is economically feasible only for the capacity range over 100 kW_{el}. For such systems expected electrical efficiency ranges between 40 and 85%, while capital and peration and maintenance (O&M) costs were estimated for 1500-3000 $kW and 0.0019 – 0.0153 $/kW, respectively [14]. By comparing these numbers with data for other systems presented in Table 2 it becomes clear that SOFC can be indeed competitive.

Vendor	Type of fuel cell	Power: electrical/ thermal [kW]	Efficiency: η_{el}/η_{tot} [%]	Number of units
Hexis	SOFC	1.0/2.0	30-35/>90	42
CFCL	SOFC	1.5/0.6	60/85	>10
Vaillant	SOFC	4.6/6.5	30/88	>60
JX Nippon	SOFC	0.7/1.25	45/87	800
Baxi	PEFC	1.0/1.7	32/85	>20
Viessmann	PEFC	2.0/5.0	28/80	10
Bosch	PEFC	4.6/6.5	29/80	10

Table 1. Comparison of selected micro-CHP systems with fuel cells.

Different concepts of large SOFC-based systems were developed and studied [15,16,17]. Among those, various plants proposed by Siemens-Westinghouse with nominal power ranging from single up to hundreds $MWel_{el}$ were investigated [18,19], including pressurized systems. Despite the fact that high efficiency and near-zero emissions in such plants were envisioned, attention has been focused on smaller scales – single and tens of kW_{el}.

Technology	Capacity	η_{el} [%]	Capital cost [$/kW]	O&M costs [$/kW]
Disel engines	500 kW - 50MW	35	200-350	0.005 - 0.015
Gas turbines	500 - 5 MW	29 - 42	450 - 870	0.005 - 0.0065
Photovoltaic system	1 kW - 1MW	6 - 19	6600	0.001 - 0.004
Wind turbines	10 kW - 2MW	25	1000	0.01

Table 2. Comparison of selected systems for stand-alone electricity generation.

3. Modeling: transition between different length and time scales

During SOFC-based system operation numbers of different processes are taking place at different length- and time-scales. Summary of typically considered phenomena is presented in Table 3.

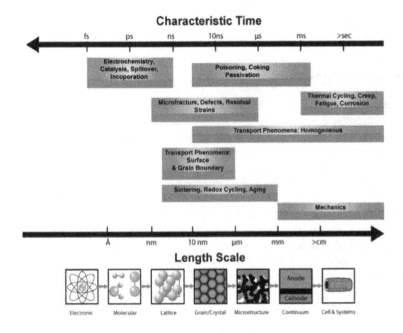

Figure 1. Different processes and their corresponding length scales and time frames (after [20])

Scale [m]	Structure	Phenomena
10^{-8}–10^{-7}	Electrode material Triple phase boundary – electrode, electrolyte and oxidant contact point	Electrochemistry Diffusion through the surface Chemical reaction
10^{-7}–10^{-5}	Porous media	Knudsen diffusion Flow through porous media Chemical reaction
10^{-5}–10^{-3}	Flow field	Diffusion Mass flow Heat exchange
10^{-3}–10^{-2}	Single cell	Transport of oxidant and fuel Thermal balancing
10^{-2}–10^{0}	Stack	Electrical circuits of the cell Processes in the electrical system Thermal balancing
10^{-0}–10^{7}	System level	Control, automatics, safety systems Integration of the entire system

Table 3. Selected processes taking place during SOFC-system operation and their corresponding length scales.

Application-specific criteria and various designs require dedicated methodology for detailed investigation of processes listed in Table 3. In general, models are used to help understand and predict behavior of a particular system, to optimize control strategy, thermal balancing, and other aspects. Additionally, optimization tools can provide information on the optimal operational parameters. Moreover, models can be used as predictive tools for performance evaluation under off-design conditions. Modeling can provide crucial information for the system configuration improvements. Work on a prototype design is usually an iterative procedure where modeling is coupled with design definition. In each case, determination of criteria is an important step and must correspond to particular requirements. Depending on type of modeling, desired complexity and level of details, sufficient data have to be supplied to model. This section will briefly review different modeling techniques, including 0, 1-2 and 3D models. In a recent and valuable summary of modeling and simulation techniques [20] pictorial illustration of different issues and their corresponding length scales (Table 3) and characteristic time has been proposed (Fig. 1).

Models can be divided into macro- and micro-scale, depending on the length scales that are covered by particular approach. In general case, analysis of SOFC at the stack level focuses on development of models for electrochemical processes, chemical reaction, transport phenomena, and geometry influence. Investigation of the entire system includes studies on the integration, heat and mass exchange, electrical circuits, and equipment.

3.1. 0D modeling

Zero-dimensional methodology allows studying processes that can be analyzed without taking into account spatial configuration and geometry. Such approach is justified for system-level studies, however might also be used for estimation of certain parameters. Depending on the required precision, 0D models can be used to solve governing equations for planar SOFC, written for each of the cell components: electrodes, electrolyte, interconnects and flow channels [21,22,23,24,25]. Required assumptions include constant fluid properties, air as an incompressible gas and no chemical reactions occurring in the fuel and air channels. Set of governing equations is later solved with desired accuracy by different algorithms. System-level studies can be performed with commercially available software such as Aspen Plus or Aspen Hysys. The later was recently used by Kupecki and Badyda [5] for evaluation of different fuel processing technologies for micro-CHP unit with SOFC. Different designs, presented in Fig. 2-6 were studied for evaluation of heat and mass balances. In the study, characteristics of market-available SOFCs were implemented, and auxiliary equipment was selected for off-the-shelf products. Considering different fuels, including natural gas, diesel, and LPG it was possible to define the optimal processing technology for micro-CHP unit equipped with afterburner. Steam reforming allowed achieving the highest overall system efficiency, even tough it required substantial amount of heat. Generally, 0D method proved to be sufficient for system-level studied, including thermal processes (i.e. heat exchange, heat losses, combustion in the afterburner), electrochemical reactions in the SOFC stack and chemical reactions occurring in the fuel processor.

With introduction of heat capacity, dynamics can be studied to some extent using 0D modeling techniques, as it will be presented in the dynamic modeling section. In certain cases chemistry can also be investigated. The main limitation is the difficulty to explicitly incorporate geometry of chemical reactors, although semi-empirical correlations are sometimes applicable. Bove and Ubertini [26] suggested using black-box 0D models to investigate impact of fuel composition, oxidant or fuel utilization and overpotentials on the macroscopic performance of SOFC in terms of efficiency and current-voltage characteristics. Such models should be used when system-level approach is required, without main focus on the SOFC stack itself [27]. In cell-leveling modeling, zero dimensional approach can be efficiently used for solving elementary balance equations for fluids: continuity, momentum, energy and species transport. Since solid oxide fuel cells consists of two porous electrodes separated by an electrolyte, porosity of these materials should be explicitly considered in the governing equations. Once the set of equations is developed, it can easily be transferred from discrete 0D to 1D model to be solved using proper CFD method [28].

Summarizing, the main advantage of zero-dimensional approach is low computational costs, simple formulation of the model. Such models can be freely used for systems where no mass and heat accumulation occurs. The main disadvantage is the significant limitation in modeling influence of geometry and sizing, especially when those are of a high importance, for instance in chemistry modeling.

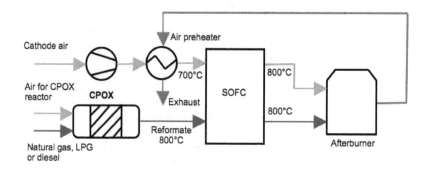

Figure 2. Single-pass system with CPOX reactor

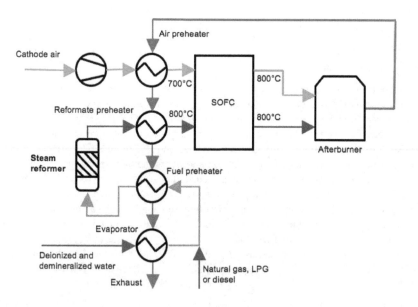

Figure 3. Single-pass system with steam reformer

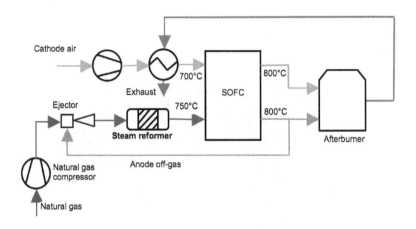

Figure 4. System with recirculation based on an ejector

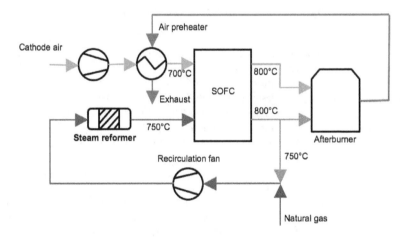

Figure 5. System with recirculation based on a high-temperature fan

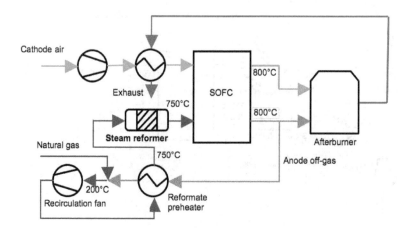

Figure 6. System with recirculation based on a low-temperature fan

3.2. 2- and 3D modeling and computational fluid dynamics codes

In 1-,2- and 3D models, space-dependent governing equations are being solved. In case of three-dimensional approach, mathematical formulas are usually written in form of partial differential equations. Different methods can be used for solving the resulting set of equations. In 1D approach, ordinary differential equations may be encountered, and solution can be easily found with simple codes or even analytically. Complex 3D models of SOFC stack are useful for heat and mass exchange modeling [29]. With high fidelity models, different heat exchange means can be studies, and cell voltage under inhomogeneous temperature distribution can be found. Space-continuous models can be applied for material studies and evaluation of process losses. Time-dependent thermal processes can be studied in similar way to proposed nearly twenty years ago by Achenbach [30]. In his work, numerical tool was used to investigate heat conductivity of stack made of ceramic and metallic plates. With the proposed methodology it was possible to find the overall heat conductivities of the combined SOFC assembly. Additionally, the model was applied to evaluate influence of thermal radiation and the total heat losses from the stack. Such studies are crucial for evaluation of overall system performance, and can indicate dangerous operational modes, which should be avoided. Several analytical models of pressure and flow distribution in the stack have been presented [59,60]. The results have been compared to 3D CFD model, showing accuracy sufficient for engineering calculations. However, analytical models are typically applicable only to no-load, isothermal stack conditions.

Significant computational power is required to implement fully-3D CFD combined models of the SOFC stack and auxiliary system components. Computational time requirements limits complex optimization of such cases. In particular, when optimization of 3D models of system sections is necessary to approximate integration of SOFC stack with pre-heaters or reformers, engineering accuracy approximations are often implemented. 3D non-CFD numerical model of SOFC stack has been applied to improve the thermal management of SOFC system through radiant heat transfer from the stack walls to adjacent air preheater panels [61].

In this study, options for minimizing axial and in-plane temperature gradients in the stack have also been identified. The results of subsequent tests, verifying modeling results, suggested that the use of radiation-based approach significantly improves the management of stack-generated heat [62].

Since porous body, representing electrochemically active part of the SOFC stack, is impermeable in directions other than flow direction in gas channels, simplifications of the 3D CFD stack model is possible, including 2D CFD model with the porous body approach (see Fig. 7). Periodical and ordered geometry of reactant channels in the stack, allows treatment of stack geometry as a porous body, with porosity defined as a ratio of channels cross-sectional area to stack cross-sectional area [63]. In the study, 2-D and 3D CFD SOFC stack models with internal manifolds have been implemented to simulate flow distribution under electric load conditions for the selected fuels. The semi-empirical model of electrochemical kinetics has been implemented. Typical flow arrangements of the inlet and outlet gas supply manifolds (U-flow, Z-flow) have been evaluated, including effects associated water-shift reaction and finite-rate of internal reforming of methane in the stack.

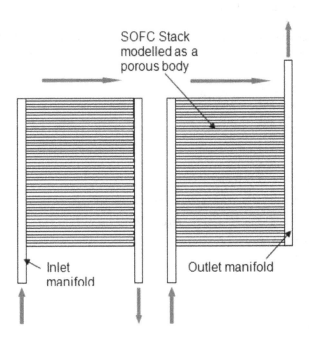

Figure 7. Stack representation in the 2D/3D CFD porous body approach

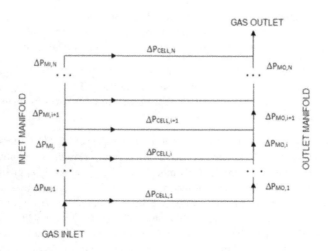

Figure 8. Stack representation in the hydraulic network approach

In yet another approach to SOFC stack modeling, so-called hydraulic network approach [64], pressure drop is calculated separately for each manifold section and reactant channel section, as shown in Fig. 8. The pressure drop is calculated based on the Darcy''s friction factor, incorporating local geometry and stream characteristics. The hydraulic model approach has been implemented for the planar, rectangular geometry of the fuel cells. In the model, pressure drop is calculated separately for each manifold section and cell section:

$$\Delta P_i = f_D \frac{L_i}{D_i} \frac{\rho_i V_i^2}{2} \tag{4.1}$$

where:

f_D Darcy's friction factor

$f_D = $ K/Re for the laminar flow and $f_D = \varepsilon/D$ for the turbulent flow

ΔP_i pressure drop in the manifold section or cell section [Pa]

L length of the manifold section or cell section [m]

ρ_i gas density [kg cm^{-1}]

V_i gas velocity [m s^{-1}]

D hydraulic diameter [m]

K constant (64 for the circular channels)

ε/D relative roughness of the channel

Re Reynolds number

Additional pressure losses are calculated for the flow obstacles, such as dividing/combining flows at the manifold/reactant channel junctions, as:

$$\Delta P_i = K_{tot} \frac{\rho_i V_i^2}{2}$$
(4.2)

The resulting system of nonlinear equations is solved numerically for each of the flow loops:

$$\Delta P_{MI,i} + \Delta P_{CELL\ ,i+1} - \Delta P_{CELL\ ,i+1} + Q \cdot \Delta P_{MO,i} = 0 \ for \ i=1...N-1$$
(4.3)

Numerical results show good convergence with analytical models (Fig. 9). The hydraulic networks approach is also applicable to SOFC stack modeling under electric load conditions.

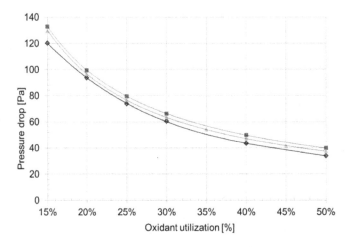

Figure 9. Comparison of pressure drop across the cathode side of the SOFC cell for a range of flows corresponding to a range of oxidant utilizations (▲– hydraulic model results; ◆ – measurements, 295 K; ■– 3D CFD simulation results)

Computational fluid dynamic system can be also used for system integration [2]. The coupling different models in one simulator can provide insight into operation of system components such as BoP devices, fuel processor, tail gas combustor, and SOFC stack. Time scale selection for the modeling should be done with caution. As has been noted by Tanaka et al. [29], certain fluctuations during small-scale co-generative SOFC-based unit operation occur. Authors performed detailed uncertainty estimation for 10 kW class units fed with town gas. Research was focused on evaluation of possible fluctuations, including changes in fuel quality over time (i.e. deviation of HHV from the nominal value), flow variations, precision of measurement equipment and other factors. The results indicate that the electrical efficiency of system can be determined with 1.0% relative uncertainty at 95% level of confidence for such system. However, it should be noted that fuel cells are generally believed to operate quite stable when compared to other energy conversion techniques [31].

3.3. Fuel processing technologies

It is well know that the main advantage of solid oxide fuel cells is the ability to operate with number of different fuels including alcohols [32], hydrocarbons [31], pure hydrogen [33,21,31], biofuels [34] and energy carries which can be converted to hydrogen-rich gas, including ammonia [35] and dimethyl ether [36]. Nonetheless, in order to assure high performance operation of a fuel cell stack, by limiting cells degradation, proper fuel processing has to be selected. As recently reported by Leone et al. [40] different fuel processing technologies may be used for fuel cell-based system, however reforming technique can influence cell operating conditions and selection should be made taking into account different factors.

Generally, three different technologies can be distinguished for converting fuel before it enters the SOFC stack: catalytic partial oxidation (CPOX), steam reforming (SR) and autothermal reforming (AT). In certain cases these processes can be accompanied by fuel clean-up stage as it is usually done for fuels with significant H_2S content.

The important part of SOFC system operation is direct thermal integration of stack and fuel reforming. Different implementations have been proposed and corresponding modeling studies performed:

1. Intermediate indirect reforming plates (IIR) can be directly integrated with the SOFC stack [45,46]. In this approach, fuel reforming plates are integrated with the stack structure and separated with one or more fuel cells. Reformed fuel from the reforming plates is redirected to fuel inlet of the adjacent cell(s).

2. Direct internal reforming (DIR) is often implemented, taking advantage of catalytic properties of the SOFC anode material [47]. In this approach fuel is directly reformed on the anode side of the fuel cell. Pre-reforming of the fuel might be necessary in some cases to avoid overcooling of the fuel inlet stack region, particularly for the fuel with high methane content.

3. Thermal integration of SOFC stack and fuel reformer can also be implemented with thermal radiation/convection/conduction conjugate heat transfer between SOFC stack(s) and reformer. In this approach, fuel reformer is placed in a direct vicinity of the stack(s).

In this subsection, theory of different fuel processing technologies will be briefly discussed.

Partial oxidation reaction proceeds with the presence of catalyst can be written in a general form for any hydrocarbon [37]

$$C_nH_mO_p + x(O_2 + 3.76N_2) + (2n-2x-p)H_2O \rightarrow nCO_2 + (2n-2x-p-0.5m)H_2 + 3.76xN_2 \quad (4.4)$$

where x is the oxygen to fuel molar ratio. This ratio defines the required amount of water for carbon to carbon monoxide conversion, amount of generated hydrogen, and molar concentration of hydrogen in the reaction products. For $x = 0$ the reaction becomes an endothermic steam reforming, and for $x = 12.5$ it corresponds to a combustion process. Partial oxidation reaction should be controlled in such way, that overall thermal balance would be exothermic. Simple calculations lead to conclusion, that for a low value of x coefficient, higher amounts (or concentrations) of hydrogen should be expected. The main reason for using catalyst is the reduction of the process temperature. Reaction described by equation (1) to proceed without catalyst, however temperature o about $1000°$ C is required in such case. Because of that fact in most commercial applications, including SOFC- based systems, catalyst is used.

Second method for turning different fuels into hydrogen-rich gas is the steam reforming. In most fuel cell applications, reaction proceeds at high temperature with addition of water vapor. Typical products of steam reforming include hydrogen and carbon dioxide. Ideal reaction can be written for any hydrocarbon fuel fed in the following form:

$$C_n H_{2n} + 0.5(n-1)H_2O \rightarrow 0.25(3n+1)CH_4 + 0.25(n-1)CO_2 \qquad (4.5)$$

In most technological processes, steam reforming comprises two stages which can be written for the simplest hydrocarbon in a form:

$$CH_4 + H_2O \rightarrow CO_2 + 3H_2 \qquad (4.6)$$

and

$$CO_2 + H_2O \rightarrow CO_2 + H_2 \qquad (4.7)$$

Where equation (4.6) is strongly endothermic and (4.7) is slightly exothermic, therefore the overall reaction requires heat delivery. Typically, steam reforming of gases can also be done as a catalyst-supported process. Usually a metallic nickel catalyst [38,39] either Ni/Al_2O_3 or Ni on refractory material, containing 5-30% of Ni are used. Lifetime of a catalyst strongly depends on quality of gases converted in the steam reformer, so-called poisoning is usually the main process leading to rapid performance deterioration. In order to ensure long lasting operation of the catalyst, poisonous impurities should be removed prior the reforming process.

Next to CPOX and SR, internal reforming is also mentioned as a interesting processing technology and the most economical way to convert hydrocarbon fuels for tubular and planar SOFCs. Despite the fact that process has number of advantages, it may lead to high temperature variations in the fuel cell and stack [41]. Highly endothermic character of the reaction is responsible for local cooling of the cell material leading to cracking and rupture. In a similar way, CPOX reaction along the cell is often claimed to be responsible for cell overheating which can compromise the ceramic material stability in a similar way as internal reforming. Even though, internal reforming is allowed to a certain extent, it is believed that thermal decomposition of higher hydrocarbons may lead to carbon formation on the anode compartments [30]. Usually, limitation on the fraction of higher hydrocarbons is imposed by material issues (endothermic reforming reaction) and possibility of carbon deposition to occur. Recent study presented in 0D modeling section and available literature [42,43] clearly indicates that steam reforming is the most efficient technology for bioethanol, methane and other hydrocarbons conversion into hydrogen rich-gas. Arteaga et al. [44] performed thorough evaluation of different fuel processing technologies particularly for SOFC application also finding steam reforming the most suitable. Comparison of different fuel processing technologies for biogas and methane was previously done and reported [53]. It was clearly indicated that steam reforming is the optimal selection for micro-CHP units with SOFCs.

As it was discussed in previous section, systems with SOFCs require fuel processing. In most cases steam reforming would be selected, and in such system catalysts would be employed.

4. Control strategy for micro-CHP unit

Since micro-CHP unit of discussed type generated both electricity and heat, two control strategies are possible. Device can operate following electricity or heat/hot water demand. Generally it is believed, that the most optimal strategy is to control electricity generation, considering heat as a by-product which can possibly be stored in sufficiently large water tank. In available literature, different tank sizes were considered. In design of micro-CHP system with power output of about 2 kW by Kupecki and Badyda [5], tank with volume of 600 liters was considered. At average storage temperature of 55° C, the total of about 28 kW_{th} can be stored in the tank. This volume was selected based on availability of off-the-shelf products, its reasonable price and sufficient heat capacity. Surprisingly, some authors [65] suggest selection of much larger size like 1000 or even 3000 liters. From the product development point of view, customer expectations and required compact size, this is an a way too large volume. Moreover, according to authors" own calculations, selection of such a big vessel has negligible economical gain, and in all cases can lead to increase of capital cost of the system. Price of hot water storage tanks increases exponentially with the capacity increase, therefore considerations of volumes above 600 liters should not have place. Additional aspect of control strategy selection depend on current conditions, including generation price and cost or resources.

In a recent study [66] control strategy for the highest energy savings for 20 households using 0.7 kW micro-CHP systems with SOFC was evaluated. Mixed-integer linear programming was used to optimized control for a case, and when each of households has a different consumption. They clearly found, that electrical load-following is the best strategy, allowing the highest energy-saving effect. Authors also pointed out that wastage of surplus hot water is possible in the summer season, but this can be avoided by selection of slightly larger water tank.

5. Modeling the dynamic behavior of a singular Solid Oxide Fuel Cell

5.1 Dynamic oriented model of SOFC

The mathematical model of SOFC for steady state calculations was presented in a few previous papers [73,74,75,76,77,78,79,80]. In this section only dynamic oriented relationships are included and commented on.

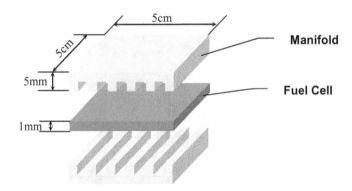

Figure 10. Dimensions of fuel cell plate and the manifolds

As an object for modeling, a singular fuel cell is chosen with dimensions of 5 cm × 5 cm and thickness of 1 mm (see Fig. 10). It was assumed that the manifolds (for fuel and oxidant) are identical.

Some processes which occur during fuel cell operation are very rapid, thus they can be assumed to be time independent compared to others. The following processes are assumed to be time independent:

1. Electrical processes

2. Electrochemical processes

3. Pressure changes

For those processes only the static equations were utilized.

Fig. 11 presents a concept of a model of Solid Oxide Fuel Cell, the fuel cell is equiped with two inlet streams and two outlet streams. Processes which occur during fuel cell operation can be divided into three steps: capture of oxygen atoms from the delivered oxidant (air), oxygen ions passing through the electrolyte layer, and the ions escaping and reacting with the delivered fuel. Material aspects play a crucial role here [68].

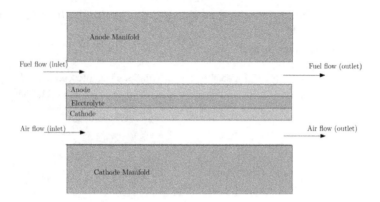

Figure 11. A concept of a model of Solid Oxide Fuel Cell

The fuel cell presented in Fig. 11 can be reduced to an 0D model. This is the simplest approach, but generates a model of the same class as models of other equipment (compressors, pumps, heat exchangers). The set of equations for the 0D model is as follows:

$$
\begin{cases}
\dfrac{dT_{Cell}}{dt} = \dfrac{\dot{Q}_{Oxidant} + \dot{Q}_{Fuel} - 2 \cdot \dot{Q}_{Surrounding} - P_{SOFC} + \dot{Q}_{Fuel}}{2 \cdot C_{Manifold} + C_{Fuel} + C_{Oxidant} + C_{Cell}} \\[4mm]
\dfrac{dp_{Cathode,Out}}{dt} = \dfrac{\dot{m}_{Cathode,In} - \dot{m}_{Cathode,Out}}{\dfrac{V_{Cathode}}{R_{Oxidant} \cdot T_{Cell}}} \\[4mm]
\dfrac{dp_{Anode,Out}}{dt} = \dfrac{\dot{m}_{Anode,In} - \dot{m}_{Anode,Out}}{\dfrac{V_{Anode}}{R_{Fuel} \cdot T_{Cell}}}
\end{cases}
\tag{6.1}
$$

where:

$$
\dot{Q}_{Oxidant} = \dot{m}_{Oxidant} \cdot c_{p,Oxidant} \left(T_{Cathode,In} - T_{Cell} \right)
\tag{6.2}
$$

$$
\dot{Q}_{Fuel} = \dot{m}_{Fuel} \cdot c_{p,Fuel} \left(T_{Anode,In} - T_{Cell} \right)
\tag{6.3}
$$

$$
\dot{Q}_{Surrounding} = k_{Surrounding} \cdot A_{Cell} \left(T_{Cell} - T_{Surrounding} \right)
\tag{6.4}
$$

$$
P_{SOFC} = E_{SOFC} \cdot I_{SOFC}
\tag{6.5}
$$

$$
\dot{Q}_{Fuel} = \dot{m}_{Fuel} \cdot HHV_{Fuel} \cdot \eta_f
\tag{6.6}
$$

$$
C_{Manifold} = m_{Manifold} \cdot c_{p,Manifold}
\tag{6.7}
$$

$$C_{Cell} = m_{Cell} \cdot c_{p,Cell} \qquad (6.8)$$

Factor	Value	Comment
Specific heat of oxidant, $c_{p,Oxidant}$ [kJ kg^{-1} K^{-1}]	1.156	Electrochemistry air
Specific heat of fuel, $c_{p,Fuel}$ [kJ kg^{-1} K^{-1}]	15.25	hydrogen
Heat transfer coefficient to surrounding, $k_{Surrounding}$ [W m^{-2}K^{-1}]	0.1	fuel cell is isolated
Higher Heating Value of fuel, HHV_{Fuel} [MJ kg^{-1}]	144	hydrogen Thermal balancing
Specific heat of interconnector material, $c_{p,Manifold}$ [kJ kg^{-1}K^{-1}]	0.5	LaCrO$_3$
Specific heat of fuel cell, $c_{p,Cell}$ [kJ kg^{-1}K^{-1}]	0.5	YSZ
Interconnector weight in relation to fuel cell area [kg m^{-2}]	20.3	
Fuel cell weight in relation to fuel cell area [kg m^{-2}]	6	

Table 4. Selected factors of a dynamic oriented mathematical model of SOFC.

The factors used in the above equations are presented in Table 4.

Typical interconnects have a thickness of 3 mm 82, which gives 7.5 cm^3 of material for fuel cell dimensions of 5 cm × 5cm × 3 mm. Assuming that the interconnect is made from La-CrO$_3$, the interconnect weight is 50 g per fuel cell. The additional weight relates to the manifolds which deliver the working fluids—depending on the current architecture solution of the stack. In this study, it was assumed that the interconnect weight in relation to fuel cell area is 2.03 g cm^{-2}.

The typical channel within which working fluids are delivered has dimensions of 0.5 mm × 1.5 mm, and its length depends on the total fuel cell dimensions (5–8 cm). Usually, the distance between the channels are the same as the channels themselves. Assuming a planar fuel cell of dimensions of 5 cm × 5 cm, the channel volume is 0.5 mm × 5 cm × 5 cm–(17 × 17 × 1.5 mm × 1.5 mm × 0.5 mm) = 0.925 cm^3 per each fuel cell side and in total 1.85 cm^3 for the fuel cell. Relating the volume to the fuel cell area gives a value of 0.074 cm^3/cm^2 of the channel volume in relation to fuel cell area.

Parameter	LaCrO$_3$
Heat conductivity [67] [W m^{-1}K^{-1}]	1.7–2.5

Coefficient of thermal expansion [72], ΔL/L/K	$(2-8)\cdot 10^{-6}$
Density [71] [g cm^{-2}]	3–6.77

Table 5. Main material parameters of interconnector.

Working fluids velocities inside the channels depend on the channel dimensions and quantity of flows delivered. To provide an adequate time for reaction as well as mixing of reagents, the velocities of working fluids should be relatively low. Based on the authors" own calculations, the nominal velocities of working fluids are below 5 m s^{-1}, being on average 1.6 m s^{-1}. Due to such low velocities, the pressure drops along the channels can be omitted [69].

Parameter	Value
Specific fuel cell weight [g cm^{-2}]	0.6
Specific interconnect weight [g cm^{-2}]	2.03
Specific volume [cm^3 cm^{-2}]	0.074

Table 6. Main material factors of the fuel cell related to fuel cell area.

5.2. Dynamic behavior of SOFC

5.2.1. The control strategy

During fuel cell operation there is a series of processes that affect its performance. The operator affects only some of them; the parameters subject to direct regulation are:

1. Temperature of inlet air

2. Temperature of inlet fuel

3. Quantity of inlet air

4. Quantity of inlet fuel

5. Electric current draw from the cell

The amounts of air and fuel supplied to the fuel cell should enable its proper operation, especially the behavior of the quantities of both fuel utilization and oxidant utilization. In addition, changes in certain parameters interact in a similar way: maintaining the desired temperature of fuel cells can be achieved by either reducing or increasing the amount of air and its temperature. Both of these parameters are related to each other (you cannot cool the cell with overly hot air, regardless of the amount). Selection of the optimal control strategy in this case is a key issue.

In this study, it was assumed that the fuel utilization factor is kept constant at the point of maximum efficiency (in fact at the laboratory scale it means only 4.5% for a fuel utilization factor of 12%). This means that inlet fuel mass flow is correlated with fuel cell current.

The most important parameter is cell temperature, which must be kept constant. The temperature is controlled by an inlet air mass-flow, which is regulated by a valve equipped with a PID regulator.

	K_p	K_I	K_D
PI	$\dfrac{1}{K \cdot T_0}$	$4.3 \cdot T_0$	
PID	$\dfrac{1 \cdot 36}{K \cdot T_0}$	$1.6 \cdot T_0$	$0.5 \cdot T_0$

Table 7. Optimal parameters of PID controller [70].

The singular PID controller is chosen to keep the fuel cell temperature at set point (800°C). The PID controls inlet air mass flow. Optimal PID parameters are listed in Table 7. For these parameters the optimal parameter settings for the PID controller of the fuel cell are determined:

- $K_p = 3.264$
- $K_I = 1.6$
- $K_D = 0.5$.

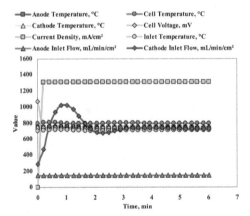

Figure 12. System response to a step change in charging current density (0–1.3 A/cm²) using PID

Fig. 12 shows the cell parameters change with a stepped increase in current density using the PID controller. It can be seen that the quality of control is very good (almost no distortion: 15°C), and the system reaches a steady state after about 4 minutes. The amount of air fed to the cathode reaches 1000 ml min^{-1}cm^{-2}. Cell voltage drops to the value of 0.75 V and remains without significant change.

5.2.2. Start-up

An external source of heat is required to support the start-up of a fuel cell. The simplest solution is to use the burner boot to warm the cell to a temperature which enables it to work independently. During fuel cell start-up acceptable temperature differences should be preserved, with the assumed values being:

• 45°C between inlet and outlet temperatures of working fluids

• 90°C between working fluid temperature and fuel cell temperature

An active start-up system is proposed, comprising regulating the temperature of gases supplied to the cells depending on cell temperature.

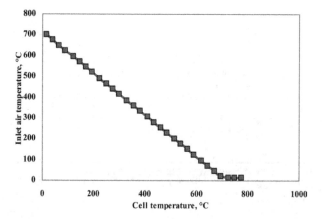

Figure 13. Correlation of air temperature used for heating the cell with cell temperature was applied during the simulation starting from cold state

The amount of air used to heat the cells was determined as the nominal point. The temperature of the air is correlated with the cell temperature according to the relation with which the air temperature decreases from the value of 700°C in proportion to the increase in cell temperature (see Fig. 13).

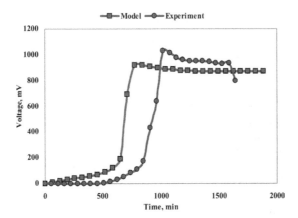

Figure 14. Comparison of the simulated start-up procedure with the results obtained from experiments [80]

Fig. 14 presents a comparison of the simulated cell voltage during start-up against the real values (data [80]). The two cells differ in structure as well as in the procedures used during start-up but, qualitatively speaking, the modeled start-up is a very close approximation to the reality.

5.2.3. Continuous operation and changes in power

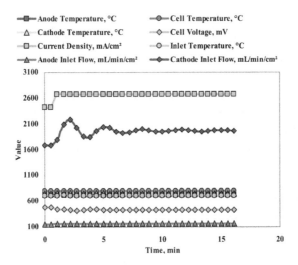

Figure 15. Changes of the fuel cell operating parameters during load increases by 10%

The results of simulated rapid increase in power by 10% are shown in Fig. 15. The control system keeps all key parameters in acceptable ranges. Larger changes are noted only for current density (which increases from 2.43–2.68 A cm^{-2}), and the air flow rate, which is the result of regulation and oscillates between 1600–2200 ml min^{-1}cm^{-2} finally stays at 2000 ml min^{-1}cm^{-2}.

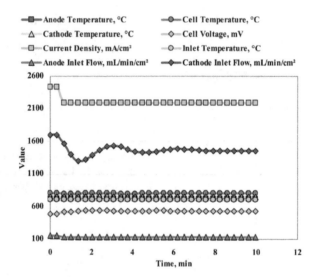

Figure 16. Changes of the fuel cell operating parameters during load decreases by 10%

The results of simulated rapid decrease in power by 10% are shown in Fig. 16. The control system keeps all crucial parameters in the acceptable ranges, so that most of the parameters practically do not change themselves. Larger changes are noted only for the current density (which increases from 2.43–2.19 A cm^{-2}), and the air flow rate which is the result of regulation and oscillates between 1300–1700 ml min^{-1}cm^{-2} finally stays at 1400 ml min^{-1}cm^{-2}.

5.2.4. Loss of load

The most likely emergency scenario is a sudden loss of load resulting from load shut down (e.g. activation of the safety switch). In this case the fuel cell should be left to idle.

Fig. 17 presents the simulated behavior of a fuel cell reacting to a sudden loss of load. The fuel cell parameters stabilize after about 4 minutes and the cell goes into idle mode. The cell temperature reaches 807°C, which seems to be a safe value.

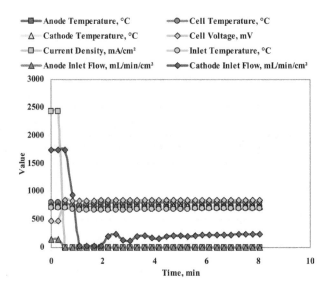

Figure 17. Simulated behavior of a fuel cell reacting to a sudden loss of load

On the basis of the simulation it can be concluded that the fuel cell is relatively resistant to a sudden loss of load in the presence of a proper control system.

5.2.5. Shut down

During normal operation the fuel cell is able to resist a sudden loss of load. Therefore no special procedures are required to shut down the fuel cell unit. Additional procedures should be used to cool down the fuel cell to ambient temperature. The best option seems to be using a PID controller with variable temperature settings.

Fig. 18 shows the fuel cell characteristics with stepped increase in the quantity of air supplied to the cathode to its maximum value (6000 ml min^{-1}cm^{-2}). The cathode part of the fuel cell loses heat relatively quickly, reaching ambient temperature after about 10 minutes. By contrast, at the anode side, there is no fuel flow and cooling takes far longer (after 30 minutes the temperature drops by only a few degrees). In total, this leads to very large temperature differences between the anode and cathode side (almost 600°C).

In order to shut down the fuel cell, the flows on both the anode and cathode sides need to be maintained. The simplest solution is to maintain the fuel stream on the anode side, which otherwise experiences a loss of fuel. Another solution is to provide another gas, but it should be an inert gas (the use of air may result in oxidation of the anode surface).

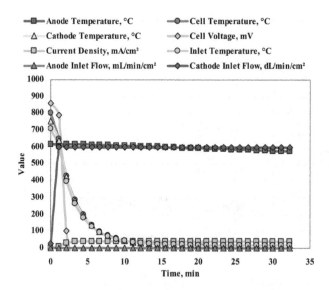

Figure 18. Fuel cell parameters during shut down procedure based on maximizing cathode flow (air side)

5.3. Discussion

The control strategy for a singular solid oxide fuel cell is proposed. The strategy is based on a singular PID controller which controls the amount of air delivered to the cathode side of the fuel cell. Additionally, fuel mass flow is correlated with current density to achieve a fixed fuel utilization factor. In fact, the efficiency of the singular laboratory scale fuel cell unit is relatively low, as is the fuel utilization factor.

The start-up procedure of the fuel cell must be supported by an external source of heat. Theoretically speaking, it is possible to heat up the cell until the point at which it starts to generate some voltage (practically, above 0.4 V) and then the fuel cell should be able to heat itself up to working temperature by the applied external load. The simulations performed do not confirm this theoretical speculation. After the load is applied, the voltage drops and no current can be drawn. Thus, adequate correlation of air inlet temperature with cell temperature is proposed in order to reach the nominal temperature. The simulated start-up was compared with the experimental data, with satisfactory results.

During normal operation, the proposed control system keeps all fuel cell parameters within acceptable ranges—there are no consequences following a rapid increase/decrease of load by 10%.

The one conceivable emergency scenario was analyzed: rapid loss of external load. The control system keeps the key parameters at acceptable levels (e.g. cell temperature reaches 807°C).

The simulated shut down procedure was unsuccessful: the PID controller was used to cool the fuel cell, but it only influenced air flow, causing an extremely high temperature gradient. Additional procedures must be applied to cool the fuel cell properly.

6. Conclusions

Different mathematical models are useful for evaluation and predicition of fuel cells and entire system performance. In all cases, specific application-related criteria are selected for development of numerical tool.

Development of advanced energy systems, including micro-CHP units, under various operating conditions is possible only with high-fidelity numerical simulator. Tool has to be validated against available experimental data. In certain cases, numerical modeling is not possible without supporting experimental measurement.

Different time- and length-scales can be covered with dedicated models, ranging from 0D up to complex full-3D tools. Steady state can be evaluated with available engineering software, including computational fluid dynamic tools.

Analysis of chemical and electrochemical reactions taking place during SOFC operation requires specific knowledge and in most cases detailed models are needed. Spatial configuration and influence of geometry can only be studied with space-continuous models or number of simplifications is required.

Based on mathematical modeling, an analysis of the dynamic operation of a singular fuel cell is presented. Based on the analysis a few cases relating to the cell were simulated:

- Start up

- Continuous operation (with power changes in the range of +/-10%)

- Shut down, and

- Emergency scenario (loss of load)

In almost all cases, the singular PID controller is able to keep the fuel cell operation within a safe range. Special procedures are required during start up and shut down. During start up, external heat sources are required to warm the cell to operational temperature. It is proposed that air temperature be correlated to cell temperature. As regards the shut down procedure, a change in fuel cell configuration is required—an inert gas instead of fuel must be delivered in order to cool the cell.

The start up procedure was compared against available experimental data with satisfactory results, qualitatively speaking.

Acknowledgements

Support from the European Regional Development Fund and Ministry of Science and Education under the project no. UDA-POIG.01.01.02-00- 016/08-00, from the National Centre for Research and Development under the project Advanced Technologies for Energy Generation, and European Social Fund through the "Didactic Development Program of the Faculty of Power and Aeronautical Engineering of the Warsaw University of Technology" are gratefully acknowledged.

Author details

Jakub Kupecki[1,2*], Janusz Jewulski[1] and Jarosław Milewski[2]

*Address all correspondence to: jakub.kupecki@ien.com.pl

1 Fuel Cell Department, Institute of Power Engineering, Poland

2 Institute of Heat Engineering, Warsaw University of Technology, Poland

References

[1] European Commission . (2004). 2004/8/EC Directive on the promotion of cogeneration based on a useful heat demand in the internal energy market and amending directive 92/62/EEC.

[2] Kattke, K. J., Braun, R. J., Colclasure, A. M., & Goldin, G. (2011). High-fidelity stack and system modeling for tubular solid oxide fuel cell system design and thermal management. *Journal of Power Sources*, 196, 3790-3802.

[3] Blum, L., Deja, R., Peters, R., & Stolten, D. (2001). Comparison of efficiencies of low, mean and high temperature fuel cell systems. *International Journal of Hydrogen Energy*, 36, 11056-11067.

[4] Mekhilef, S., Saidur, R., & Safari, A. (2012). Comparative study of different fuel cell technologies. *Renewable and Sustainable Energy Reviews*, 16, 981-989.

[5] Kupecki, J., & Badyda, K. (2011). SOFC-based micro-CHP system as an example of efficient power generation unit. *Archives of Thermodynamics*, 32(3), 33-42.

[6] DOE Energy Efficiency and Renewable Energy Information Center. (2008). Comparison of fuel cell technologies.

[7] Mai, A., Sfeir, J., & Schuler, A. (2007). Status of sofc stack and systems development at Hexis. Fuel Cell Seminar 2007 proceedings 5-8.11. 12.

[8]   Schuler, A., Nerlich, V., Doerk, T., & Mai, A. (2010). Galileo 1000N–status of development and operation experiences. *Proceedings of 9th European Solid Oxide Fuel Cell Forum*, 2, 98.

[9]   Foger, K. (2010). Commercialisation of CFCL"s residential power station Blugen. *Proceedings of European Fuel Cell Forum, Lucerne*, 2, 22-29.

[10] Brennstoffzellenheizgerate fur die Hausenergieversorgung: Die Zukunft der Kraft-Warme- Kopplung. (2008). (2036197-204), VDI-Bericht.

[11] Klose, P. (2009). Pre-series fuel-cell-based m-CHP units in their field test phase. In 11th Grove Fuel Cell Symposium. London.

[12] Pawlik, J., & Klaschinsky, H. (2008). Das Viessmann Brennstoffzellen- Heizgerat. (2036189-196), VDI-Bericht.

[13] Toshiyuki Unno JX Nippon. personal communication.

[14] Kirubakaran, A., Jain, S., & Nema, R. K. (2009). A review of fuel cell technologies and power electronic interface. *Renewable and Sustainable Energy Reviews*, 13, 2430-2440.

[15] Dunbar, W. R., Lior, N., & Gaggioli, R. A. (1991). Combining Fuel Cells with Fuel-Fired Power Plants for Improved Exergy Efficiency. *Energy*, 16(4), 1259-1274.

[16] Bedringas, K. W. (1997). Exergy Analysis of Solid-Oxide Fuel Cell (SOFC) Systems. *Energy*, 22(4), 403-412.

[17] Chan, S. H., Ho, H. K., & Tian, Y. (2002). Modelling of simple hybrid solid oxide fuel cell and gas turbine power plant. *Journal of Power Sources*, 109, 111-120.

[18] Harvey, S. P., & Richter, H. J. (1997). Gas turbine cycles with solid oxide fuel cells. Part I: improved gas turbine power plant efficiency by use of recycled exhaust gases and fuel cell technology. *Journal of Energy Resources Technology*, 116, 305-311.

[19] Kupecki, J. (2010). Integrated Gasification SOFC Hybrid Power System Modeling: Novel numerical approach to modeling of advanced power systems. VDM Verlag Dr. Müller.

[20] Grew, K. N., & Chiu, W. K. S. (2012). A review of modeling and simulation techniques across the length scales for the solid oxide fuel cell. *Journal of Power Sources*, 199, 1-13.

[21] O"Hayre, R., Cha, S. W., Colella, W., & Prinz, F. (2005). Fuel cell fundamentals. Wiley.

[22] Pasaogullari, U., & Wangm, C. (2003). Computational fluid dynamics modeling of solid oxide fuel cells. *Electrochemical Society Proceedings*, 07, 1403-1412.

[23] Proceedings of hydrogen and fuel cells 2004, Toronto, Canada. Mathematical modeling of the transport phenomena and the chemical/electrochemical reactions in solid oxide fuel cells: A review. (2004).

[24] Dagan, G. (1989). Flow and transport in porous formations. Springer.

[25] Iwata, M., Hikosaka, T., Morita, M., Iwanari, T., Ito, K., & Onda, K. (2000). Performance analysis of planar-type unit sofc considering current and temperature distributions. *Solid State Ionics*, 132, 297-308.

[26] Bove, R., & Ubertini, S. (2006). Modeling solid oxide fuel cell operation: Approaches, techniques and results. *Journal of Power Sources*, 159, 543-559.

[27] Lisbona, P., Corradetti, A., Bove, R., & Lunghi, P. (2007). Analysis of a solid oxide fuel cell system for combined heat and power applications under non-nominal conditions. *Electrochimica Acta*, 53, 1920-1930.

[28] Karcz, M. (2009). From 0D to 1D modeling of tubular solid oxide fuel cell. *Energy Conversion and Management*, 50, 2307-2315.

[29] Tanaka, T., Inui, Y., Urata, A., & Kanno, T. (2007). Three dimensional analysis of planar solid oxide fuel cell stack considering radiation. *Energy Conversion and Management*, 48, 1491-1498.

[30] Achenbach, E. (1994). Three-dimensional and time-dependent simulation of a planar solid oxide fuel cell stack. *Journal of Power Sources*, 49, 333-348.

[31] US Department of Energy, Office of Fossil Energy NationalEnergyTechnologyLaboratory. Fuel Cell Handbook 7th Edition. (2004). *EG G Technical Services, Inc.*

[32] Jiang, Y., & Virkar, A. V. (2001). A high performance, anode-supported solid oxide fuel cell ope- rating on direct alcohol. *Journal of Electrochemical Society*, 148(7), A706-A709.

[33] Yokokawa, H. (2003). Understanding materials compatibility. *Annual Review of Materials Research*, 33, 581-610.

[34] Staniforth, J., & Ormerod, R. M. (2003). Running solid oxide fuel cells on biogas. *Ionics*, 9(5-6), 336-341.

[35] Wojcik, A., Middleton, H., Damopoulos, I., & Van Heerle, J. (2003). Ammonia as a fuel in solid oxide fuel cells. *Journal of Power Sources*, 118(1-2), 342-348.

[36] Murray, E. P., Harris, S. J., & Jen, H. (2002). Solid oxide fuel cells utilizing dimethyl ether fuel. *Journal of Electrochemical Society*, 149(9), A1127-A1131.

[37] Ahmend, S., Krumpelt, M., Kumar, R., Lee, S. H. D., Carter, J. D., Wilkenhoener, R., & Marshall, C. (1998). Catalytic partial oxidation reforming of hydrocarbon fuels. Technical Report CMT/CP96059, Argonne National Laboratory.

[38] Simell, P., Kurkela, E., Stahlberg, P., & Hepola, J. (1996). Catalytic hot gas cleaning of gasification gas. *Catalysis Today*, 27, 55-62.

[39] Ma, L., Verelst, H., & Baron, G.V. (2005). Integrated high temperature gas cleaning: tar removal in biomass gasification with a catalytic filter. *Catalysis Today*, 105, 729-734.

[40] Leone, P., Lanzini, A., Delhomme, B., Ortigoza-Villalba, G. A., Smeacetto, F., & Santarelli, M. (2012). Analysis of the thermal field of a seal-less planar solid oxide fuel cell. *Journal of Power Sources*, 204, 100-105.

[41] Morel, B., Roberge, R., Savoie, S., Napporn, T. W., & Meunier, M. (2007). An experimental evaluation of the temperature gradient in solid oxide fuel cells. *Electrochemical and Solid-State Letter*, 10(2), B31-B33.

[42] Wang, Y., Yoshiba, F., Kawase, M., & Watanabe, T. (2009). Performance and effective kinetic model of methane steam reforming over Ni/YSZ anode of planar SOFC. *International Journal of Hydrogen Energy*, 34, 3885-3893.

[43] Ivanov, P. (2007). Thermodynamic modeling of the power plant based on the sofc with internal steam reforming of methane. *Electrochimica Acta*, 52(12), 3921-3928.

[44] Arteaga, L. E., Peralta, L. M., Kafarov, V., Casas, Y., & Gonzales, R. (2008). Bioethanol steam reforming for ecological syngs and electricity production using a fuel cell SOFC system. *Chemical Engineering Journal*, 136(2-3), 256-266.

[45] Dokmaingama, P. (2010). Modeling of IT-SOFC with indirect internal reforming operation fueled by methane: Effect of oxygen adding as autothermal reforming. *International Journal of Hydrogen Energy*, 35, 13271-13279.

[46] Dokmaingama, P. (2009). Modeling of SOFC with indirect internal reforming operation: Comparison of conventional packed-bed and catalytic coated-wall internal reformer. *International Journal of Hydrogen Energy*, 34, 410-421.

[47] Ioraa, P., Aguiarb, P., Adjimanb, C. S., & Brandonb, N. P. (2005). Comparison of two IT DIR-SOFC models: Impact of variable thermodynamic, physical, and flow properties. Steady-state and dynamic analysis. *Chemical Engineering Science*, 60, 2963-2975.

[48] Churakova, E. M., Badmaev, S. D., Snytnikov, P. V., Gubanov, A. I., Filatov, E. Y., Plyusnin, P. E., Belyaev, V. D., Korenev, S.V., & Sobyanin, V. A. (2010). Bimetallic rheco/zro2 catalysts for ethanol steam conversion into hydrogen- containing gas. *Kinetics and Catalysis*, 51, 893-897.

[49] Yang, Y., Ma, J., & Wu, F. (2006). Production of hydrogen by steam reforming of ethanol over a ni/zno catalyst. *International Journal of Hydrogen Energy*, 31, 877-882.

[50] Snytnikov, P. V., Badmaev, S. D., Volkova, G. G., Potemkin, D. I., Zyryanova, M. M., Belyaev, V. D., & Sobyanin, V. A. (2012). Catalysts for hydrogen production in a multifuel processor by methanol, dimethyl ether and bioethanol steam reforming for fuel cell applications. *International Journal of Hydrogen Energy*.

[51] Sobyanin, V. A., Cavallaro, S., & Freni, S. (2002). Dimethyl ether steam reforming to feed molten carbonate fuel cells (MCFCs). *Energy Fuels*, 14, 1139-1142.

[52] Fleisch, T. H., Sills, R. A., & Briscoe, M. D. (2002). Emergence of the gas-to liquids industry: a review of global gtl developments. *Journal of Natural Gas Chemistry*, 11, 1-14.

[53] Kupecki, J., Jewulski, J., & Badyda, K. (2011). Selection of a fuel processing method for SOFC-based micro-CHP system. *Rynek Energii*, 97(6), 157-162.

[54] Faungnawakij, K., Tanaka, Y., Shimoda, N., Fukunaga, T., Kawashima, S., Kikuchi, R., & Eguchi, K. (2007). Hydrogen production from dimethyl ether steam reforming over composite catalysts of copper ferrite spinel and alumina. *Applied Catalysis B: Environmental*, 74, 144-151.

[55] Yamada, Y., Mathew, T., Ueda, A., Hiroshi, S., & Tetsuhiko, K. (2006). A novel DME steam- reforming catalyst designed with fact database on demand. *Applied Surface Science*, 252, 2593-2597.

[56] Tanaka, Y., Kikuchi, R., Takeguchi, T., & Eguchi, K. (2005). Steam reforming of dimethyl ether over composite catalysts of g-Al2O3 and Cu-based spinel. *Applied Catalysis B: Environmental*, 57, 211-222.

[57] Semelsberger, T. A., Ott, K. C., Borup, R. L., & Greene, H. L. (2006). Generating hydrogen-rich fuel-cell feeds from dimethyl ether (DME) using physical mixtures of a com- mercial Cu/Zn/Al2O3 catalyst and several solid-acid catalysts. *Applied Catalysis A: General*, 65, 291-300.

[58] Feng, D., Wang, Y., Wang, D., & Wang, J. (2009). Steam reforming of dimethyl ether over CuO-ZnO-Al2O3-ZrO2 + ZSM-5: A kinetic study. *Chemical Engineering Journal*, 146, 477-485.

[59] Kee, R. J., Korada, P., Walters, K., & Pavol, M. (2002). A generalized model of the flow distribution in channel networks of planar fuel cells. *Journal of Power Sources*, 109(1), 148-159.

[60] Maharudrayya, S., Jayanti, S., & Deshpande, A. P. (2006). Pressure drop and flow distribution in multiple parallel-channel configurations used in proton-exchange membranes fuel cell stack. *Journal of Power Sources*, 157(1), 358-367.

[61] Krist, K., & Jewulski, J. (2006). A Radiation-Based Approach to the Design of Planar, Solid Oxide Fuel Cell Modules. *Journal of Materials Engineering and Performance*, 15, 468-473.

[62] Jewulski, J., Krist, K., Petri, R., & Pondo, J. (2008). Gas Process Panels Integrated with Solid Oxide Fuel Cell Stacks, U.S. Patent No. 7374834.

[63] Jewulski, J., Blesznowski, M., & Stepien, M. (2009). Flow Distribution Analysis of the Solid Oxide Fuel Cell Stack under Electric Load Conditions, Proceedings of the 9th European SOFC Forum, Lucerne, Switzerland.

[64] Boersma, R. J., & Sammes, N. M. (1997). Distribution of gas flow in internally manifolded solid oxide fuel-cell stacks. *Journal of Power Sources*, 66, 441-452.

[65] Alanne, Kari, Saari, Arto, & Ugursal, V. Ismet. (2006). Joel Good The financial viability of an SOFC cogeneration system in single-family dwellings. *Journal of Power Sources*, 158, 403-416.

[66] Wakui, T., Yokoyama, R., & Shimizu, K. (2010). Suitable operational strategy for power interchange operation using multiple residential SOFC (solid oxide fuel cell) cogeneration systems. *Energy*, 35, 740-750.

[67] Badwal, S., Deller, R., Foger, K., Ramprakash, Y., & Zhang, J. (1997). Interaction between chromia forming alloy interconnects and air electrode of solid oxide fuel cells. *Solid State Ionics*, 99(3-4), 297-310.

[68] Andrade, T., & Muccillo, R. (2011). Effect of zinc oxide and boron oxide addition on the properties of yttrium-doped barium zirconate. *Ceramica*, 57(342), 244-253.

[69] Christman, K., & Jensen, M. (2011). Solid oxide fuel cell performance with cross-flow roughness. *Journal of Fuel Cell Science and Technology*, 2011, 8(2), 024501.1-5.

[70] Findeisen, W. (1973). Poradnik inzyniera automatyka. Wydawnictwa Naukowo-Techniczne.

[71] Furusaki, A., Konno, H., & Furuichi, R. (1995). Pyrolitic process of la(iii)-cr(vi) precursor for the perovskitc type lanthanum chromium oxide. *Thermochimica Acta*, 253, 253-264.

[72] Hayashi, H., Watanabe, M., & Inaba, H. (2000). Measurement of thermal expansion coefficient of lacro3. *Thermochimica Acta*, 359(1), 77-85.

[73] Milewski, J. (2010). Advanced mathematical model of sofc for system optimization. In: ASME Turbo Expo 2010: Power for Land, Sea and Air. No. GT2010-22031. ASME.

[74] Milewski, J. Advanced model of solid oxide fuel cell. *Fuel Cell Science, Engineering & Technology Conference. No. FuelCell 2010-33042*, ASME.

[75] Milewski, J. (2010). Mathematical model of SOFC for complex fuel compositions. *International Colloquium on Environmentally Preferred Advanced Power Generation*, ICE-PAG2010-3422.

[76] Milewski, J. (2010). Simultaneously modelling the influence of thermal-flow and architecture parameters on solid oxide fuel cell voltage. *ASME Fuel Cell Science and Technology*.

[77] Milewski, J. (2011). SOFC hybrid system optimization using an advanced model of fuel cell. *Proceedings of the 2011 Mechanical Engineering Annual Conference on Sustainable Research and Innovation*, 121-129.

[78] Milewski, J., Badyda, K., & Miller, A. (2010). Modelling the influence of fuel composition on solid oxide fuel cell by using the advanced mathematical model. *Rynek Energii*, 88(3), 159-163.

[79] Milewski, J., & Miller, A. (2004). Mathematical model of SOFC (Solid Oxide Fuel Cell) for power plant simulations. *ASME Turbo Expo*, 7, 495-501.

[80] Milewski, J., Świrski, K., Santarelli, M., & Leone, P. (March 2011). Advanced Methods of Solid Oxide Fuel Cell Modeling. 1st Edition, Springer-Verlag London Ltd.

[81] Sakai, N., & Stolen, S. (1995). Heat capacity and thermodynamic properties of lanthanum(iii) chromate(iii): LaCr03, at temperatures from 298.15 k. evaluation of the thermal conductivity. *The Journal of Chemical Thermodynamics*, 27(5), 493-506.

[82] Zhai, H., Guan, W., Li, Z., Xu, C., & Wang, W. (2008). Research on performance of LSM coating on interconnect materials for SOFCs. *Journal of the Korean Ceramic Society*, 45(12), 777-781.

Permissions

The contributors of this book come from diverse backgrounds, making this book a truly international effort. This book will bring forth new frontiers with its revolutionizing research information and detailed analysis of the nascent developments around the world.

We would like to thank Cumhur Aydinalp, for lending his expertise to make the book truly unique. He has played a crucial role in the development of this book. Without his invaluable contribution this book wouldn't have been possible. He has made vital efforts to compile up to date information on the varied aspects of this subject to make this book a valuable addition to the collection of many professionals and students.

This book was conceptualized with the vision of imparting up-to-date information and advanced data in this field. To ensure the same, a matchless editorial board was set up. Every individual on the board went through rigorous rounds of assessment to prove their worth. After which they invested a large part of their time researching and compiling the most relevant data for our readers. Conferences and sessions were held from time to time between the editorial board and the contributing authors to present the data in the most comprehensible form. The editorial team has worked tirelessly to provide valuable and valid information to help people across the globe.

Every chapter published in this book has been scrutinized by our experts. Their significance has been extensively debated. The topics covered herein carry significant findings which will fuel the growth of the discipline. They may even be implemented as practical applications or may be referred to as a beginning point for another development. Chapters in this book were first published by InTech; hereby published with permission under the Creative Commons Attribution License or equivalent.

The editorial board has been involved in producing this book since its inception. They have spent rigorous hours researching and exploring the diverse topics which have resulted in the successful publishing of this book. They have passed on their knowledge of decades through this book. To expedite this challenging task, the publisher supported the team at every step. A small team of assistant editors was also appointed to further simplify the editing procedure and attain best results for the readers.

Our editorial team has been hand-picked from every corner of the world. Their multi-ethnicity adds dynamic inputs to the discussions which result in innovative outcomes. These outcomes are then further discussed with the researchers and contributors who give their valuable feedback and opinion regarding the same. The feedback is then collaborated with the researches and they are edited in a comprehensive manner to aid the understanding of the subject.

Apart from the editorial board, the designing team has also invested a significant amount of their time in understanding the subject and creating the most relevant covers. They scrutinized every image to scout for the most suitable representation of the subject and create an appropriate cover for the book.

The publishing team has been involved in this book since its early stages. They were actively engaged in every process, be it collecting the data, connecting with the contributors or procuring relevant information. The team has been an ardent support to the editorial, designing and production team. Their endless efforts to recruit the best for this project, has resulted in the accomplishment of this book. They are a veteran in the field of academics and their pool of knowledge is as vast as their experience in printing. Their expertise and guidance has proved useful at every step. Their uncompromising quality standards have made this book an exceptional effort. Their encouragement from time to time has been an inspiration for everyone.

The publisher and the editorial board hope that this book will prove to be a valuable piece of knowledge for researchers, students, practitioners and scholars across the globe.

List of Contributors

Guohong Wu, Yutaka Yoshida and Tamotsu Minakawa
Dept. of Electrical of Electrical Engineering & Information TechnologyTohoku Gakuin University, Japan
NPO ECA, Japan

Murat Ozturk
Department of Physics, Faculty of Art and Sciences, Suleyman Demirel University, 32260, Isparta, Turkey

G. Sevjidsuren, E. Uyanga, B. Bumaa, E. Temujin, P. Altantsog and D. Sangaa
Department of Material Science and Nanotechnology, Institute of Physics and Technology, Mongolian Academy of Sciences, Mongolia

Janusz Jewulski
Fuel Cell Department, Institute of Power Engineering, Poland

Jarosław Milewski
Institute of Heat Engineering, Warsaw University of Technology, Poland

Jakub Kupecki
Fuel Cell Department, Institute of Power Engineering, Poland Institute of Heat Engineering, Warsaw University of Technology, Poland

Printed in the USA
CPSIA information can be obtained
at www.ICGtesting.com
JSHW011319221024
72173JS00003B/33